Sustainable Agricultural Development

Sustainable Agricultural Development

Dr. K.A. Rasure

Oxford Book Company
Jaipur ▪ New Delhi

Published By

Oxford Book Company

Regd. Off. "Bony Residency", Gate No. 2
Opp. Tilak Public School, Vishweswariya Nagar,
Gopalpura Road, Jaipur - 302018, India
Ph: 0141 - 2761280, Telefax: 2761381
email - oxfordbookjpr@gmail.com
website - oxfordbookcompany.com

Delhi Office: 102, 1st Floor, Satyam House,
4327/3, Ansari Road, Darya Ganj,
New Delhi - 110002, India
Ph: 011-45652440

Distributed by
KEYA INTERNATIONAL PRESS LLP
E-mail: keyainternationalpressllp@gmail.com

Published in India, 2025

ISBN : 9789380179643

Typeset by Shivangi Computers, Jaipur

Printed and bound in India by Replika Press Pvt. Ltd.

Foreword

"Sustainable Agricultural Development" is the second UGC sponsored Two Day National Seminar conducted by our faculty of Economics. We are happy over the fact that the faculty has chosen a topic of significance to the agricultural community at large. "Sustainable Development" seeks to meet the needs and aspirations of the present, without compromising the ability to meet those of the future. The proposition that, development is an imperative necessity for a poor country like India, that the pattern of development should be sustainable and that the desired pace of development cannot be achieved without commensurate infrastructure, has been made relevant and of general acceptance.

Environment and agriculture have an age old and permanent relationship with each other. Perhaps no other human effort has interacted with environment as intimately as agriculture did. Yet the concern for environmental degradation is of recent origin. Quite a lot has happened in agriculture in this century that has contributed its share of ecological disorders. Again, it is being felt that, new agricultural technologies using an increased dose of inorganic intents have adversely affected nature, and that traditional agriculture has been neutral to it and its operations. Mankind is closely linked with environment, and its very exists is likely to be threatened if natural disorders go beyond a point, should sound be enough a warning to all of us to direct our keen attention and concern problems of environment and the measures to be adopted to rectify the disorders that have already taken place and to prevent future is disequilibrium taking place. It is in this respect that, the present National Seminar on Sustainable Agricultural Development has assumed greater importance.

The National Seminar will be addressed these several issues and truly provides an opportunity to a number of teachers, agri-scientists, researchers, farmers to express their views on the three themes, they are Environment and Sustainable Agricultural Development, Sustainable Agriculture and Allied Sectors and Strategies of Sustainable Agricultural Development. The papers will indeed be stimulating. The publication of the selected papers, we hope will receive the attention they deserve from academicians and with intelligent and create fruitful discussion. We are grateful to the faculty of economics and other staff of our college, who have rendered fullest support in this academic endeavour.

Manne Hasreddy

Preface

"Sustainable Development implies the satisfaction of the needs of the present without compromising the ability of future generations to meet their own needs". Sustainable Agricultural Development has been recognized as the right goal of economic development. Building sustainability has to be approached as a threefold task-Social, Economic and Ecological, simultaneously. The "Sustainable Agricultural Development", visualizes to restructure the traditional approach to agriculture development in a way, that the agricultural development activities are to be integrated with the environmental policies. The increasing population, lopsided developmental activities and rapid industrialization all over the world have caused ecological imbalance at local as well as global level. Thus, Sustainable Agricultural Development without environmental consideration is ruining the human life by deteriorating the life support system.

Agricultural development should not be at the cost of the environment. Neglect of environmental consideration can cause serious environmental damage and endanger the lives of both the present and future generations. The title of the book "Sustainable Agricultural Development", highlights the major environment concerns related to agriculture. It examines the sustainable agricultural and allied activities, and then it suggests strategies and other policy measures. A critical examination is made on the effectiveness of the measures taken to ensure sustainable agricultural development.

This book contain 29 papers prepared by eminent personalities coming fairly from seats of higher learning situated in different parts of India. Each chapter of this book is well written by the leading experts having long standing in the research in their respective fields. The book occupies special advantage of excellence because it is an edited version and ingests the best refined ideas of eminent agri-scientists, teachers, and research scholars from different disciplines.

At the outset, we are highly grateful to Deputy Secretary, UGC, SWRO, Bangalore for the liberal financial support, and all those, who have contributed financially.

We are grateful to our beloved President, Sri. Baswaraj Patil, Sedam, Secretary, Dr. Nagreddy Patil, Joint Secretary, Sri. Govindlal Malpani, Treasurer, Sri. Ramanujadas Tapadia and all the executive committee members of Sri. Kottal Basweshwar Bharatiya Shikshan Samiti, for their encouragement and cooperation and who even interest in every aspect of the National Seminar.

We thank our Principal Sri. Manne Hasreddy who as chairman of the organizing committee has given needed encouragement and cooperation.

We are thankful to all our colleagues and non-teaching staff for their involvement, hard work, and encouragement.

We are especially grateful to the learned contributors for their scholarly write-ups. The present book is our humble contribution to the cause. Hopefully, it will shed some light to take us out of this grim situation.

Lastly, we express our gratitude to M/s Oxford Book Company, Jaipur, for publishing it so impressively.

Dr. K.A. Rasure

List of Contributors

A. Marimuthu
*Lecturer in Economics,
R.D.Govt.Arts.College, Sivagangai-630561,
Tamilnadu*

A. Sakthivel
*Lecturer in Economics,
Selvamm Arts and Science College,
Namakkal, Tamilnadu*

A.R. Hundekar
*University of Agricultural Sciences, Dharwad,
AINRP (T), Agricultural Research Station,
Karnataka,
Nipani-591 237.*

B.G. Lobo
*Lecturer of Economics, Sinhgad Institute, N.B.
Navale College of commerce, Kusgaon (Bk),
Lonavala, Pune-410401*

B.P. Veerabhadrappa
*Reader and Head, DOS in Economics,
P.G. Centre, Shiva-gangothri,
Kuvempu University, Davangere-02,
Karnataka.*

B.V. Patil
*Special Officer, University of Agricultural
Sciences, Raichur -584102,
Kamataka, India.*

G.N. Kulkarni
*Department of Soil Science and Agricultural
Chemistry, College of Agriculture, Dharwad.*

Bachinamani Monappa
*Lecturer in Economics,
Govt. Arts Grade College,
Guirmitkal - 585 214 (Dist) Gulbarga.*

D. Pulla Rao
*Department of Economics,
Andhra University Visakhapatnam, A.P.*

Girish . G.
*Krishi Vigyan Kendra, Aland Road,
Gulbarga, University of Agricultural Sciences,
Dharward.*

Gouradevi Katnalli
*Selection Grade Lecturer in Economics,
V.G. Womens College, Gulbarga.*

H. Venkatesh
*AICRP for Dryland Agriculture,
Bijapur Centre, University of Agricultural
Sciences, Dharwad.*

H.B. Babalad
*Senior Scientist (Agronomy), Institute of
Organic Farming, UAS, Dharwad- 5.*

J.B. Gopali
Agricultural Research Station, Gulbarga 585101, University of Agricultural Sciences, Raichur, Kamataka, India.

Jagannath, B. Kukkudi
Faculty, Department of Management, Gulbarga University, Gulbarga - 585 106.

Jaganath, M. Lagashetty
Lecturer in Economics, Mahatma Jyotiba Phule Degree College, Shahabazar, Gulbarga.

Jennifer Serto
Research Scholar.

K. Banadarangaiah
Professor in Commerce, Acharya Institute of Management and Sciences, Bangalore.

K. Meenatchi Sundaram
Final Year Student in M.A. Economics, R.D. Govt. Arts. College, Sivagangai-630561, Tamilnadu.

K.A. Rasure
Selection Grade Lecturer in Economics, Nrupatunga Arts & Commerce College, Dist. Gulbarga (Karnataka) Sedam-585 222.

M.B. Guled
AICRP for Dryland Agriculture, Bijapur Centre, University of Agricultural Sciences, Dharwad.

M.K. Biradar
SGL in Economics MSI Degree College Gulbarga.

M.S. Shirahatti
AICRP for Dryland Agriculture, Bijapur Centre, University of Agricultural Sciences, Dharwad.

M.S.Kallur
Retd.Professor of Economics, Gulbarga University, Gulbarga.

M.S. Shirahatti
AICRP for Dryland Agriculture, Bijapur Centre, University of Agricultural Sciences, Dharwad.

M.V. Manjunath
AICRP for Dryland Agriculture, Bijapur Centre, University of Agricultural Sciences, Dharwad.

M.V. Ranghswami
Tamilnadu Agricultural University.

Mamatha, B.
Selection Grade Lecturer, Smt. V.H.D Central Institute of Home Science, Bangalore.

Muniswamy
Krishi Vigyan Kendra, Aland Road, Gulbarga, University of Agricultural Sciences, Dharawad.

P. Loganathan
Reader in Economics, Kandaswami Kandar's College, P. Velur, Namakkal, Tamilnadu.

P.S. Dharmaraj
ARS, Gulbarga, University of Agricultural Sciences, Raichur - 584 102.

P.S. Matiwade
University of Agricultural Sciences, Dharwad, AINRP (T), Agricultural Research Station, Karnataka, Nipani-591 237.

Panduranga Venkataramulu
Assistant Professor, Acharya Institute of Management & Sciences, Bangalore – 58.

Pavitra Ravindra Alur
Ph.D. Scholar, Dept. of Social Work, Solapur University, Solapur

R. Sivasamy
Tamilnadu Agricultural University.

R.N. Padaria
Senior Scientist (Agricultural Extension), Indian Agricultural Research Institute, New Delhi-110012

Revanasiddeshwara, M. Kintimath
Selection Grade Lecturer in Economics, Govt. First Grade College, Hesaraghatta, Bangalore - 560 089.

Rupendra, R. Gaikwad
Assistant Professor at Shri Shivaji Maratha Society's Institute of Management & Research, Pune-09.

S. Nalinakshi
H.O.D of Economics, Vijaya Degree College, R.V. Road, Basavanagudi, Bangalore – 560 004.

S. Santana Bosu
Tamilnadu Agricultural University.

S.A. Patil
Director, Indian Agricultural Research Institute, New Delhi-110012

S.K. Dhage
Reader and H.O.D. of Economics, Baburaoji Gholap College, Sangvi, Pune - 27.

S.M. Matapati
Krishi Vigyan Kendra, Aland Road, Gulbarga, University of Agricultural Sciences, Dharawad.

S.R. Mulla
AICRP for Dryland Agriculture, Bijapur Centre, University of Agricultural Sciences, Dharwad.

S.T. Hundekar
Department of Soil Science and Agricultural Chemistry, College of Agriculture, Dharwad.

S.T. Kajjidoni
University of Agricultural Sciences, Dharwad, AINRP (T), Agricultural Research Station, Karnataka, Nipani-591 237.

S.V. Santhosh
Research Scholar, DOS in Economics, P.G. Centre, Shiva-gangothri, Kuvempu University, Davangere-02, Karnataka.

Sherly Kurien
Reader, Department of Economics, Avinashilingam University for Women, Coimbatore - 641043.

Raju. G. Teggelli
Krishi Vigyan Kendra, Aland Road, Gulbarga, University of Agricultural Sciences, Dharawad.

T. Gurubasappa
Selection Grade Lecturer, P.G. Dept. of Economics, Govt. College, Gulbarga.

V. Basil Hans
HOD of Economics & Dean, Faculty of Arts, St Aloysius Evening College, Mangalore – 575 003, D.K. District, Karnataka, India.

V.B. Kuligod
Department of Soil Science and Agricultural Chemistry, College of Agriculture, Dharwad.

Vivekanand, A. Mane
Lecturer in Commerce, Sinhgad Institutes, N.B. Navale College of Commerce, Lonavala, Pune-10.

Yamanura
Krishi Vigyan Kendra, Aland Road, Gulbarga, University of Agricultural Sciences, Dharawad.

Contents

1

Sustainable Agriculture—Principles and Praxiology

S.A. Patil and R.N. Padaria

Introduction

The concept and philosophy of sustainable agriculture assumed global significance aftermath the adverse social and environmental impact of modern agriculture. Spearheaded with use of high yielding varieties, external inputs and non-renewable energy; agricultural modernization was adopted as essential means for development and transformation of rural communities and nation's prosperity. However, the euphoria of its potential benefits began to wane on account of regional imbalances and social inequity in development and rapid environmental degradation. Massive and injudicious application of synthetic fertilizers and toxic pesticides besides large-scale irrigation led to colossal damage to ecosystem and more importantly the human and cattle health. Disquieting impacts include (Pretty, 1996):

1. Contamination of water by pesticide, nitrates, and wastes, causing harm to wildlife, disruption of ecosystem and possible health problem.
2. Contamination of food and fodder by pesticide residues.
3. Damage to farm and natural resources by pesticides.
4. Contamination of atmosphere by ammonia, methane etc.
5. Formation of ozone depletion, global warming and atmospheric pollution.
6. Over use of natural resources causing depletion of ground water.
7. Displacement of traditional varieties and breeds by modern varieties/ breeds.
8. New health hazards due to agrochemicals—during field spray and working in manufacturing industries.

Injudicious and rampant use of chemical fertilizers has led to imbalance in soil nutrient status and thus food grains are devoid of essential nutrients leading to increasing levels of hidden hunger. Similarly the injudicious and rampant use of pesticides has resulted in the problem of pesticidal residues in blood as well as mothers' milk. Such effects have put the human and cattle in trap of deadly diseases and have also threatened the existence of other living beings. Realizing the catastrophic impact of pesticides, the famous American writer Carl Rachel in her famous book "The Silent Spring" has warned that human being is an integral part of nature and any war against nature is the war against self. She further advocates that often the technological advancements become so antagonistic with the nature's activities and system that it is wise to lay them off.

However, the proponents of modern and exploitative agriculture still argue in their favour and advocate their continuance for food security and prosperity.

The debate is mounting between modernists and environmentalists and conservationists over issue related to sustainability and food security.

While ensuring food security to burgeoning population is the prime concern, conservation of natural resources and protection of environment for deriving benefits in perpetuity are equally important. Consequently vision and action towards balanced and sustainable development were recognized and accepted by the global community of scientists, policy makers and development agencies.

Concepts and Philosophy

The word sustain, which has been derived from Latin word root "*sustinere*" (*sus* – meaning from below and *tenere* meaning to hold), means to keep existence or maintain implying long-term support or permanence.

Though the concept of "sustainability" first came up during the energy crisis of 1970s in relation to supply of oil, when it was defined as "maintaining the present without compromising the future" (Jennings and Ikerd, 1994), the term "sustainable development" first appeared in the document 'The World Conservation Strategy" published in 1980 by the International Union for the Conservation of Nature and Natural Resources (Workman, 2005), and the notion was been defined as an "integration of preservation (nature protection) and development to provide a planetary change which can ensure safe survival and welfare of all people" (Krapivin and Varotsos, 2007). The principle of uniform distribution of resources, which was the cornerstone of Brundtland Commission Report, and democratic participation, which was emphasized by the Second U.N.Conference on the Environment and Development in Rio de Janeiro are the important features for sustainability.

However, different societies have different conceptualization of sustainability as well as different requirements of sustainability based upon varying cultural expectations or environmental constraints (Brown *et. al,* 1987). According to Bruntland (1987), the core of the idea of sustainability is the concept that current decisions should not damage the prospects for maintaining or improving living standards in future.

With mounting criticism and growing recognition of problems with modernists' approach to agriculture as well as increased public awareness about environmental issues, search for conceptualization, description and operationalization of alternative forms of agriculture intensified. Sustainable agriculture emerged as the term to describe the varied field of agricultural practices that differ from conventional concepts of modern agricultural production (Hauptli, *et.al.* 1990).

Keller (2002) cites the observation of the highly renowned agricultural scientist and thinker, Dr.M.S.Swaminathan, that in a world where 30 million infants were born with mental impairment due to low birth weight and Third World countries like India had over 250 million people living below the poverty line, sustainability in agriculturewas no longer a choice, which was a necessity. According to Swaminathan (1999) the success of sustainable agriculture results from a combination of science, technology, service and public policy. He further quotes "Varro", a Roman landowner as the earliest proponent of sustainable agriculture, who said in the First century B.C. : "Agriculture is a science which teaches us what crops should be planted in each kind of soil, and what operations are to be carried on, in order that the land may produce the highest yields in perpetuity." Sustainable agricultural systems are those, which are capable of maintaining their productivity and usefulness to society indefinitely, and such systems must be resource conserving, socially supportive, commercially competitive and environmentally sound (Duesterhans, 1990).

Sustainable Agriculture has been defined by Food, Agriculture, Conservation and trade Act (FACTA, 1990) as an integrated system of plant and animal production practices having a site specific application that will, over the long term

- Satisfy human food and fiber needs;
- Enhance environmental quality;
- Promote efficient use of nonrenewable resources;
- Sustain the economic vitality of farm operations; and
- Enhance the quality of life for farmers and society as a whole.

Source: U.S Congress, 1990; Horne and McDermott (2005)

Leopold Center in Iowa has defined sustainable agriculture as "farming systems that are environmentally sound, profitable, productive, and maintain the social fabric of the rural community."

Many view sustainable agriculture as philosophy based on co-existence and communion with nature, where emphasis is upon understanding of long-term impact of our activities on the environment and on other species.

Brown (1984) defines sustainable agriculture as stewardship. It emphasizes stewardship of both natural and human resources. This includes concern over the living and working condition of farm laborers, consumer health and safety, and the needs of rural communities.

In attempt to describe the alternative forms of agriculture in contrast to modern agriculture many terms were used viz., sustainable agriculture, regenerative agriculture, eco-agriculture, low external input agriculture, low input sustainable agriculture, resource—conserving agriculture, organic agriculture, permaculture, etc., which led to confusion and misconception. The most common misconception about sustainable agriculture has been that it represents a return to some form of low technology or backward or traditional agricultural practices, which is untrue. It envisages utilization of economically and ecologically viable old as well as new innovations. Sustainable agriculture does not mandate a specific set of agricultural practices, rather a basket of options and approaches according to need could be tried. However, the basic challenge with sustainable agriculture remains how best to optimally utilize the internal resources.

Goals for Sustainable Agriculture

According to Pretty (1996) a sustainable agriculture is any system of food or fibre production that systematically pursues the following goals:

1. Thorough incorporation of natural processes such as nutrient cycling, nitrogen fixation and pest-predator relationship into agricultural production process.
2. A reduction in the use of off-farm, external and non-renewable inputs.
3. A more equitable access to productive resources and opportunities, and progress towards more socially-just forms of agriculture.
4. Productive use of the biological and genetic potential of plant and animal species.
5. An increase in self-reliance among farmers and rural people.
6. Long-term sustainability of current production levels.

7. Profitable and efficient production with an emphasis on integrated farm management, and the conservation of soil, water, energy and biological resources.

Dimensions and Parameters of Sustainable Agriculture

In order to practice sustainability, it is imperative to have its holistic understanding. Swaminathan (1991) identified 14 major dimensions of sustainable agriculture covering the social, economical, technological, political and environmental facets of sustainability as technological appropriability, economic feasibility, economic viability, environmental soundness, temporal stability, resource-use-efficiency, local adaptability, social acceptability, political tacitness, administrative manageability, cultural desirability, equity, and productivity. Lockeretz (1988) delineated following physical and biological parameters for sustainable agriculture

1. Diversity of crop species.
2. Selection of crops and livestock that are adapted to particular environment.
3. Preference for farm generated resources rather than purchased materials.
4. Tightening of nutrient cycles to minimize nutrient losses.
5. Livestock housed and grazed at low stocking densities.
6. Enhancement of storage of nutrient in the soil.
7. Maintenance of protective cover on the soil.
8. Rotation that include deep rooted crops and control weeds.
9. Use of soluble inorganic fertilizer.
10. Use of pesticide for crops protection only as a last resort.

CUESA'S Guiding Principles of Sustainable Agricultural Production

Centre for Urban Education about Sustainable Agriculture, San Francisco (2005) envisions a healthy world nourished by sustainable food systems. The guiding principles for sustainable agricultural production advocated by them stresses upon environmentally sound, humane, economically viable and socially just technologies:

Environmentally Sound: Producers actively work to create and sustain cultivated landscapes that are complex, diverse and balanced biological systems. Producers use practices that conserve and restore resources.

Humane Animal Management: While being raised, animals are allowed to engage in the natural behaviors that are important to their well being, and are harvested in ways that minimize stress to the animals and the environment.

Economically Viable: Producers operate within a framework of sound business planning and pursue integrated and proactive approaches to marketing and sales.

Socially Just: Producers and their employees receive fair and reasonable compensation and work in a safe and respectful environment.

Sustainable Agriculture: Pragmatic Roadmap

Transition to sustainable agriculture cann't be an instantaneous act. Farmers and practitioners first need to have conviction about the potential of sustainable agricultural practices in meeting the ever-growing demands of food, feed, fibre and fuel of the families and community at large. Already the society across the world is engulfed in debate whether food sufficiency could be maintained with sustainable agricultural practices or it will revert to subsistence level. However, there are remarkable evidences to prove that with sustainable agricultural practices not only the level of productivity could be maintained but also the production bases could be replenished and rejuvenated.

A pragmatic roadmap to sustainable agriculture requires integrated emphasis upon adoption of resource-conserving technologies, participation of farmers and their collectives, and partnership and support of political and service organizations. Besides these elements, what are of paramount importance are value-orientation and perception of practitioners towards sustainable agriculture. For way to sustainable agriculture, Krapivin and Varotsos (2007) have emphasized upon conservation and replenishment of natural resources (soil and water), use of eco-friendly methods of pest management, conservation of energy resources, conservation of biodiversity, recycling of organic waste and location specific crops and breeds of animals (Box-2).

STEPS TO SUSTAINABLE AGRICULTURE
1. Create and conserve healthy soil.
2. Conserve water and protect its quality.
3. Manage organic wastes without pollution.
4. Manage pests with minimal environmental impact.
5. Select livestock and crops adapted to the natural environment.
6. Encourage biodiversity.
7. Conserve energy resources.
8. Increase profitability and reduce risk.

Source: (Krapivin and Varotsos, 2007)

Shift in Value-orientation and Perception of Practitioners and Policy-makers

Values are the guiding forces of human behaviour. For facilitating transition to sustainable agriculture it is imperative to begin from the level of change in value-orientation as well as the perception of individuals. Brown, *et.al.*(1987) highlighted that values were the key to the evolution of a sustainable society. For augmentation of sustainable agriculture it is imperative to learn to live in communion with nature. Wisdom, values and methods to co-exist with nature are deeply ingrained in traditional agricultural systems. We need to take some leaves out of the treasure of indigenous knowledge and integrate them with modern practices to have sustainability in the process. Documentation, validation, integration and scaling up of indigenous technologies and practices are called for standardization of sustainable agricultural practices.

Augmentation of Systems Thinking

Farming has to be viewed in system perspective with adequate focus upon interrelatedness and interdependence of important constituents or elements of the whole system. Generally a rural system is viewed to possess five basic assets: natural capital, social capital, human capital, physical capital and financial capital (Fig. 1); which are transformed by processes, institutions and policies to deliver outcomes to the society in form of deliverables for existence and sustenance. A sustainable system is one, which accumulates and increases these bases overtime and one, which depletes its capital base, is called as an unsustainable system. Therefore, for sustainability to prevail, it is desired that activities are designed

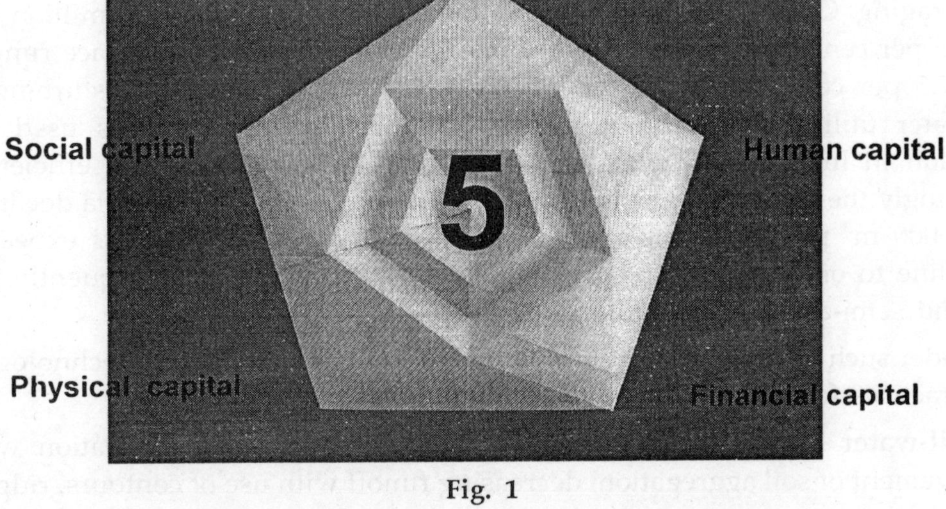

Fig. 1

and performed in system perspective i.e. taking into consideration the impact and consequences of one activity upon other related activities.

Adoption of Resource-conserving Technologies and Their Impact

Zealous pursuit of higher productivity undoubtedly led to spectacular increase in food grain production and self-sufficiency but sadly at relatively higher social and environmental costs. One of the most critical disadvantages has been the disappearance of integrated and mixed farming systems. Agricultural intensification led to high degree of specialization of crop cultivation systems leading to over-reliance upon external inputs, overexploitation of groundwater, ecological pollution, degradation of land and water bodies,and health hazards. However, the damage to the production system was more appalling on account of the disconnect of crop and livestock enterprises, which in traditional system were the integral components and helped in rejuvenating the system.

The principles of integrated farming are keys to sustainable agriculture and thus a variety of technologies and practices for pest, nutrient, soil and water management is required for practicing sustainable agriculture.

Management of Soil and Water Resources

India has diverse forms of soil with varying land quality (capability) classes. About 125 Mha of land areas (under land quality classes of II to VI) are prone to soil related constraints including drought stress, high temperatures, and nutrient imbalance; which need sustainable management through adoption of recommended management practices for higher water use efficiency (WUE) and increasing crop yields (Rattan, 2008). The scenario for water resources is equally discouraging. Of the total annual rainfall received in India i.e., 400 million ha-m; 37.5 per cent infiltrates into soil, while 45 per cent is lost as surface run-off and 17.5 per cent as evaporation (Bhaskar, 2002). What is more disturbing is the water utilization efficiency. About 200 km^3 water per year is used for irrigation in India but astonishingly with bare 20-30 per cent use efficiency. Alarmingly the per capita availability of fresh water resources in India declined from 6000 m^3 per year in 1947 to 2260 m^3 per year in 1997 and is expected to decline to only 1130 m^3 per year by 2047 (Rattan, 2008). Consequently the arid and semi-arid regions would witness severe water shortages.

Under such circumstances it is essential to promote adoption of technologies and practices for soil and water conservation.

Soil-water Management: Principles of increasing water infiltration with improvement of soil aggregation, decreasing runoff with use of contours, ridges, vegetative hedges and reducing soil evaporation with use of crop residues mulch could be employed for better management of soil-water.

Soil surface management: Technologies like deep tillage, conservation tillage, mulch farming, etc., could be used for increasing soil-water retention and reducing soil evaporation.

Soil conservation practices
1. Conservation tillage
2. Contour farming
3. Mulches and cover crops
4. Silt traps and gully fields

Several studies have shown the beneficial impact of such practices. Laryea *et.al* (1991) observed yield increase in chick–pea, maize and sorghum in Andhra Pradesh to the tune of 231, 29 and 28 per cent, respectively by deep ploughing. Increase in yield of sorghum by about 34 per cent was observed by Patil and Sheelavantar (2006) in Karnataka and Selva Raju *et al.* (1999) in Tamil Nadu. Similarly there was highly appreciating increase in yield of rainfed crops viz., maize and sorghum due to mulching (Table 1). Practices like contour cultivation, and cultivation on graded bunds help in effective conservation of moisture. These practices reduce runoff upto 40 per cent and contribute to yield increments upto 25-35 per cent depending upon the rainfall situation (Singh and

Table 1

Increase in yield of rainfed crops due to mulching

Crop	State	% increase by Mulching	Reference
Maize	A.P.	14-48	Cogle et al. (1997)
	Punjab	9-44	Prihar et al. (1979)
	Punjab	25-29	Ghuman and Sur (2001)
	Himachal	16-22	Acharya and Sharma (1994)
	Himachal	8-150	Sharma and Acharya (2000)
Sorghum	Central India	67-77	Randhawa and Venkateswarlu (1980)
	Andhra Pradesh	27-48	Cogle *et al.* (1997)

Source: Rattan (2008)

Venkateswarlu, 2009). About 20-22 per cent of cropping season rainfall in western Rajasthan, which used be lost through surface run-off and nearly 1.5 to 2 tonnes of soil per ha lost due to water erosion, could be reduced to 14-16 per cent and 0.7 -1.1 tonnes/ha with grass barriers of *Ververia zizanioides* (Khas)/ *Sachhram munja/Cenchrus ciliaris/Dichanthium annulatum* (Trivedi *et.al.*,2009)

These potential benefits amply prove the healthy and profitable impacts of soil and water conservation technologies.

Checklist for Farmers for Water Conservation

Horne and McDermott (2005) have elucidated following checklist of practices for farmers to conserve water and protect its quality:

1. Conservation tillage methods.
2. Increasing soil organic matter.
3. Reducing chemical use.
4. Using the least toxic chemical for the job.
5. Establishing conservation buffer strips.
6. Protecting riparian areas.
7. Terracing.
8. Farming on the contour.
9. Strip cropping.
10. Building ponds to catch sediment.
11. Rotational grazing.
12. Practicing efficient irrigation.
13. Growing crops adapted to the climate, soil type and topography.
14. Fencing cattle out of water ways and ponds.

Soil fertility management: Soils are not only thirsty but hungry too. However, rampant use of chemical fertilizers in exploitative farming system led to severe ill effects to soil. Sustainable soil nutrient management philosophy advocates integrated plant nutrient management through legume – led crop rotation, use of compost, manures and biofertilizers, choice of appropriate cropping systems, etc. Green manuring helped in fertilizer N recovery by first crop to the tune of 3.5 per cent and retention of 2.4 per cent more N in soil and reduced unaccounted –for N by 6.1 per cent (Gosh et.al, 2007). Intercropping systems involving cereal and legume crops are common in dry lands (Singh et.al.1990) as means of risk adjustment, profit generation and fertility management.

With inclusion of legumes in intercropping systems,, about 0.746 million tones of N could be added (Saraf, Shinde, and Hegde,1990).

Biofertilizers are potential tools for sustainable soil fertility management. Application of *Rhizobium, Azatobacter, Azospirillum* and Phosphate solubilizing bacterial inoculants could help in saving chemical fertilizers by 15-20 kg/ ha (Table-2)

Table 2
Benefits of Biofertilizers

Biofertilizer	Amount of Application (Kg/Ha)	Input Cost (Rs)	Impact Kg N	Total Benefit (Rs)	Anticipated Benefits/ha
Rhizobium	0.5	25	20.0	190-225 540-880	19-22 Kg/ha (grain legumes) 50-80 kg/ha fodder)
Azatobacter	0.5	25	17.5	145-200	15-20 kg/ha
Azospirillum	0.5	25	20.0	200	20 kg/ha
Phosphate solubilizer	0.5	25	27.5	600	25 kg P_2O_5
Blue Green Algae	1.0	80	25.0	260-315	25-30 kg/ha
Arbuscular Mycorrhiza inoculant	1.0	80	25.0	420-670	20-30 kg P_2O_5

Dhar, *et.al.* (2009).

Adoption of Integrated Pest Management (IPM) Practices

Pesticides are not the perfect answer to controlling pests and pathogens though they account for 10-40 per cent of world's gross agricultural production, argue Conway and Pretty (1991). Deleterious effects of chemical pesticides on human health and ecosystem have forced agriculturists, even the proponents of modern agricultural technology, to look for integration of cultural practices; plant products based formulations and bio-agents for lessening the quantum of deadly pesticides. Of late there is increased acceptance of integrated pest management strategy and practices. Farmers' Field School programme got momentum to make farmers aware of pest dynamics and their eco-friendly management.

Management and control of pests and disease with use of resistant varieties and breeds; alternative natural pesticides; bacterial and viral pesticides; pheromones for disrupting pest reproduction, etc could be adopted for sustainability of production process. Bio-agents have crucial role to play in pest management.

Hence practices to promote natural enemies like release of predators and parasites; improving the habitat for natural enemies; facilitating beetle banks and flowering strips; crop rotation and multiple cropping should be integrated in pest management practices.

Non- Pesticidal Management (NPM): NPM reflects the farmers' creativity in dire adversity. Immensely plagued by debt trap due to heavy reliance upon pesticides and losing their source of livelihood, farmers in Andhra Pradesh have taken up plant product based formulations and eco-friendly strategy to pest management.

Non-Pesticidal Management of Insect pests is a "system that maintains the pest populations at levels below those causing economic injury, by having healthy crop and managing the population dynamics in the crop ecosystem" (Centre for Sustainable Agriculture, 2004).

The main principles underlying the Non-Pesticidal Management are:

1. A natural ecological balance will ensure that pests do not reach a critical number in the field that endangers the yield
2. Nature can restore such a balance if it is not meddled with too much, hence no chemical pesticides/pesticide are used in crops at all.
3. Understanding the insect biology and crop ecology is important to manage pests – it is not enough if reactive sprays are taken up during outbreak.
4. Prevention rather than control/reaction is the key element to NPM
5. Crop diversity and soil health play an important role in pest management
6. That pest management is possible with local, natural material

Non-pesticidal management uses many different practices:

1. Deep ploughing in the summer to expose the insect pupae so they dry in the sun.
2. Using light traps and bonfires to attract moths.
3. Placing yellow and white sticky boards in the field to attract insects that suck out the plant's juices.
4. Removing by hand the leaves on which many insect eggs have been laid.
5. Setting pheromone traps to check on the numbers of pests in the field.
6. Using biological pesticides such as neem seed-kernel extracts and chilli–garlic extracts to control bollworms and sucking insects. There are also other locally available plants to make biological pesticides.

7. Using an extract made from cow dung and urine to control aphids and leafhoppers, which also acts as a fertilizer.
8. Planting trap crops such as castor and marigold. Insects are likely to lay their eggs on these plants, where they can be picked off easily.

Mahesh and Rao (2008) reported successful use of non-pesticidal management of pests and farmers could get their mortgaged land back from moneylenders by the earned profit.

Challenges and Way Ahead for Sustainable Agriculture

Shifting to sustainable agriculture paradigm wouldnot be a smooth affair for the farmers. Though sustainable practices have been the domain of Indian farming since centuries but many a practice and value have gone into oblivion. It would require deskilling as well as reskilling and retooling the farmers to eschew the expoilitative practices and values to adopt conservative and evolving values and practices. Certainly the cost involved in transition and time required to learn new skills will have to be considered as a policy matter to facilitate promotion of sustainable agriculture at a wider scale. Amidst the challenging situations there are highly encouraging efforts of farmers in initiation of sustainable agiculture. Prabhu (2009) reported an excellent example of Mr. Arunachalm of Tamil Nadu who adopted multi-seeding practice in banana cultivation and earned rupees six lakhs from three acres land in 365 days without spending much on inputs. Lyla Bavadam (2005) reports the movement for sutainable agriculture gaining momentum in Maharashtra and cites Kisan Mehta of Prakurti, who argues that with modern farming techniques one would earn Rs 100 but would spend Rs. 80 on labour, buying chemicals and fuel, among other things; while in sustainable farming, one would earn Rs 80 but that entire amount would be his and he could put it back into the farm. Such successful experiences provide direction to other farmers and hence need to be replicated and disseminated at large scale for wider diffusion and adoption.

Policy Interventions for Institutional Support for Promoting Sustainable Agriculture

Operationalization of sustainable agriculture necessitates collective action, for which local institutions need to be strengthened and policy interventions be made for their adequate capacity building. Some of the important policy initiatives suggested include:

1. Establishment of a national strategy for IPM.
2. Prioritization of research with emphasis upon sustainable agriculture without undermining national interest.

3. Greater emphasis upon on-farm research and generation of appropriate technology in farming system perspective.
4. Encouraging participatory process in generation, dissemination and adoption of sustainable practices.
5. Strengthening local groups for community Action.
6. Promotion of farmers-to-farmer communication.
7. Institutionalizing support systems for sustainable agriculture practising farmers.
8. Direct subsidies and grants towards sustainable technologies.
9. Promoting public awareness programme on sustainability issues.
10. Establishing appropriate standards and regulations for pesticides.
11. Strengthening the capacities of NGOs to scale Up sustainable and indigenous practices.
12. Fostering stronger NGO-government partnership.
13. Institutional mechanism for teaching and training in sustainable practices.

The best pragmatic step towards sustainable agriculture would to be self-awakening and self-preparedness. An East Anglian proverb cited by George Ewart Evans, 1966 (Pretty (1996) reflects the true spirit for sustainable agriculture:

"A farmer should live as though he were to die tomorrow; he should but farm as though he were going to live forever"

References

Acharya, C.L. and P.D. Sharma.1994 Tillage and mulch effects on soil physical environment, root growth, nutrient uptake and yield of maize and wheat on Alfisol in north west India. *Soil and Tillage Research,* 32.

Bhaskar, K.S. 2002. Watershed concept and yield optimization under dryland farming. In A.M. Dhopte (ed.). Agrotechnology for dryland farming. Scientific Publishers, Jodhpur, India: 27-60.

Brown, G.E. 1984. Stewardship in agriculture. In Douglass, G.K.(ed) Agricultural sustainability in a changing world order. Westview Press, Boulder, Colorado.

Brown, B.J. et.al. 1987. Global sustainability: towards definition. *Environmental Management.* 11(6), 713-719.

Brundtland, G. 1987. Our Common Future: The World Commission on Environment and Development. Oxford University Press. UK.

Cogle, *et.al. 1997.* Soil management options for Alfisols in the semi-arid tropics: annual and perennial crop production. *Soil and Tillage Research.* 44.

Dhar, D. Lata, Sunil Pabbi, K. Annapurna, Geeta Singh, Sangeeta Paul, Livleen Shukla, Y.B.Sharma and M.S.Rathi.2009. Biofertilizers for better soil health and crop prodcutivity. Technical bulletin (TB-ICN: 62/2009), IARI, New Delhi

Ghosh, P.K., K.K. Bandhopadhyay, R.H. Wanjari, M.C. Manna, A.K. Misra, M. Mohanty, A. Subba Rao, 2007. Legume effect for enhancing productivity and nutrient use-efficiency in major cropping systems—An Indian perspective: A Review, *Journal of Sustainable Agriculture,* Vol. 30 (1) 2007

Ghuman, B.S. and H.S.Sur. 2001. Tillage and residue management effects on soil properties and yields of rainfed maize and wheat in in a sub-humid sub-tropical climate._*Soil and Tillage Research.* 58

Hauptli, H.*et.al.* 1990. Biotechnology and crop breeding for sustainable agriculture. In Edwards, et.al. *Sustainable Agricultural Systems.* Soil and Water Conservation Society.

Horne,J.E. and M.NcDermott.2005. The Next Green Revolution- Essential steps to a healthy, sustainable agriculture.Atlantic Publishers. New York.

Jennings, V.M. and J. Ikerd. 1994. National perspective on agricultural sustainability. Paer presented at the American Society of Agronomy North central Branch Meeting in conjuction with REAP Workshop. Des.Moines, IA, August, 1994.

Keller.2002.Geneticist talks sustainability. *The Stanford Daily.* May 17.

Krapivin, V. and Costas.A. Varotsos.2007. Globalisation and sustainable developemnt-Environmental agendas.Praxis Publishing, UK.

Lal, Rattan, 2008. Managing soil water to improve rainfed agriculture in India, *Journal of Sustainable Agriculture,* Vol. 32 (1) 2008.

Laryea, K.B.*et.al.1991.* Tillage systems and soils of the semi-arid tropics. *Soil and Tillage Research.* 20: 201-218.

Lockeretz, W. 1988. Open questions in sustainable agriculture. *American Journal of Alternative Agriculture.* 3 (4), 174-181.

Lyla, Bavadam.2005. A green alternative. *Frontline.* Volume 22. July 16-29, 2005.

Mahesh and DUMRao.2008. Non-pesticidal management in crops: community managed extension, processes and impacts. Unpublished M.Sc. Thesis. IARI, New Delhi

Parihar, S.S. *et.al.* 1979. Effect of mulching previous crops or fallow on dryland maize and wheat. *Experimental Agriculture,* 15. 129-134.

Patil, S.L. and M.N. Sheelavantar.2006. Soil water conservation and yield of winter sorghum (Sorghum bicolor L. moench) as influenced by tillage, organic materials and nitrigen fertilizer in semi-arid tropical India. *Soil and Tillage Research,* 89:246-257.

Prabhu, M.J. 2009. Sustainable farming: low investment – maximumreturns. *The Hindu.* January 1, 2009.

Pretty, J.N.1996. Regenerating agriculture-Policies and practice for sustainability and self-reliance.Vikas Publishing House Pvt.Ltd., New Delhi.

Randhawa, N.S. and J.Venketaswarlu.1980. Indian experience in the semi-arid tropics- prospects and retrospects. ICRISAT Proceedings of the International Symposium on Developemnt and Transfer of Technology for rainfed Agriculture and SAT Framers. 28 August-1 sept, 1979. ICRISAT, Hyderabad.

Saraf, C.S., V.S. Shinde, and R., Hegde. 1990. Agronomic research towards sustainble agriculture. Indian Society of Agronomy, New Delhi, pp. 153-162.

Selvaraju, R., P. Subbian, A. Balasubramaniana, R.Lal. 1999. Land configuration and soil nutrient management options for sustainable crop production on Alfisols and Vertisols of southern peninsular India. *Soil and Tillage Research.* 52: 203-216.

Sharma, P.K. and C.L.Acharya.2000. Carry over of residual soil moisture with mulching and conservation tillage practices for sowing of rainfed wheat in northwest India. *Soil and Tillage Research*, 57, 43-52.

Singh, A.K., B. Venkateswarlu, 2009. Conservation Agriculture in Arid and Semi-arid regions: Opportunities and Constraints ,Indian Farming, February 2009.

Singh, R.P. *et.al*. 1990. Field manual on watershed management, CRIDA, Hyderabad, pp. 165.

Swaminathan, M.S. 1991. Towards sustainable agriculture: dimensions and components. The Hindu Survey of Indian Agriculture.

Swaminathan, M.S. 1999. New Agriculturist – *On-line*, 99/4

Trivedi, T.P. 2009. *Indian Farming*. Vol58. No.11. 2009, p. 8.

Workman, J.2005. Pragmatic Solution: an assessment of progress. Atar Roto Press, Genewa, Switzerland, pp. 80.

2

Environment and Sustainable Agriculture

H.B. Babalad

Globally, the achievements of modern science based agriculture can hardly be over stated. Producers have readily adopted a succession of mechanical, biological and chemical innovations that have transformed agriculture into a powerful industrial machine that produces abundant food. The exemplary performance during late sixties, mid seventies and early eighties helped to over come the food insecurity.

However the heavy dependence of modern agriculture on non-renewable fossil fuel based synthetic fertilizers, pesticides and energy needs of cultivation, harvest, intensive animal production and grain processing, raise the questions about the long term sustainability of agriculture. Also the full cost of chemical and energy intensive agriculture on degradation of natural resources and quality of air and water environment is largely debited against the gain in productivity.

Agriculture Scenario

Food security has been major area of concern for planners as well as scientists in India since long food grain production in India has made impressive gains from 50.22 million tones immediately after independence (1950-51) to 230.7 million tones today (2007-08) which is suppose to be self sufficiency. The present self sufficiency in food production on one hand did not ensure food security of individuals. As the food security means every individual has the physical, economic, Social and environmental access to a balanced diet that includes necessary macro and micro nutrients, safe drinking water, sanitation, environmental hygiene, primary health care and education so as to lead a healthy and productive life.

Agriculture has grown at an annual rate of 3.4 per cent during 1980's but it declined to 1.8 per cent during late 90's and during early 20th century it further reduced to 1.5 per cent. The productivity of most of the crops have stagnated at about 1980's level with little growth beyond that. The farmable growth was recorded only in case of wheat, rice and a slight growth in ragi.

Diversification of agriculture in order to meet the changing demand was adopted as one of the important strategies to cope up with the volatile prices through deliberate policy changes and encouragement of private participation. Diversification was in forour of high value support oriented crops. According to a recent study (Joshi *et al.*, 2004), agricultural diversification increased in all South Asian countries. Fruits and Vegetable have shown good performance during 1980's and 1990's. Non food grain crops have gradually replaced food grain crops. The shift is more towards non food grain crops like oilseeds, fruits, vegetables, spices and cotton. In India, Southern region is highly diversified fallowed by western region.

Employment in agricultural sector registered a negative trend. Employment on usual status basis has registered a growth rate of 1.54 per cent during 83-93 but showed a meager growth rate 0.18 per cent during 90's. The formers gross and net income is not increasing at the rate at which it grew in the years following green revolution. The fertilizers prices too grew at a very fast rate and net income of the farmers could not keep pace with it. In this context is the small and marginal farmers who are caught in a vicious circle of low investment, low productivity and low income.

Intensive Agriculture and Natural Resources

Soil degradation processes such as desertification, erosion, decline in soil organic matter, soil contamination (e.g., by heavy metals), soil compaction, decline in soil bio-diversity and salinisation can cause soil to lose its capacity to carry out its main functions. Such degradation processes can result from inappropriate farming practices such as unbalanced fertilization, improper use of irrigation water, over – abstraction of groundwater for irrigation, improper use of pesticides, use of heavy machinery or overgrazing. Soil degradation may also result from abandonment of certain farming practices. For example greater specialization towards arable farming has frequently been accompanied by abandonment of traditional crop rotation systems and manuring with green legumes and the practices that contributed to the restoration of soil organic matter content.

To maintain the productivity of 1980's we started using more and more fertilizers. It has increased from 0.52 kg/ha during 1950-51 to 96.4 kg/ha of NPK. It varied from state to state and it is 196 kg/ha in Punjab, 164 kg/ha in Haryana and 10 kg/ha in states like Arunachal Pradesh, Nagaland and Sikkim (Anonymous, 2008). It is estimated that consumption of chemical fertilizers increased seven folds and pesticides by 375 times, while the food production has just double during the first 20 years of the launch of the green revolution in India (Ramanathan, 2006). Though high input intensive technologies have increased agriculture production and labour efficiencies these appear to have an adverse effect on soil productivity and environment qualities on long run. The use ratio of NPK fertilizers is also not balanced which is 7.9:3:1 as against 4:2:1. The imbalanced use of chemical fertilizers and the absence of organics was noticed in intensive farming. The other practices monocropping and poor irrigation management practices have triggered the ill effects of intensive farming.

During green revolution the cropping intensity and per unit productivity increased with introduction and usage of high yielding varieties and hybrids, chemical fertilizers and irrigation. In the process of this transformation the crops and varieties choosen were selective and high yielding due to lack of diversity that were found prone to epidemic pest and diseases. This led to extensive use of pesticides.

Groundwater Contamination

According to some estimates, groundwater contamination accounts for nearly 80 per cent of the domestic water needs, and 50 per cent of the urban water needs in India. Groundwater is generally less susceptible to contamination and pollution when compared to surface water bodies. Also, the natural impurities in rainwater which replenishes groundwater system, get removed while infiltrating through soil strata. But, In India, where groundwater is used intensively for irrigation and industrial purposes, a variety of land water-based human activities are causing pollution of this precious resource. Its over-exploitation is causing aquifer contamination in certain instances, while in contamination in certain instances, while in certain other its unscientific development with insufficient knowledge of groundwater flow dynamic and geo-hydro chemical processes has led to its mineralization.

Nitrate concentration is above the permissible level of 45 ppm in 11 states, covering 95 districts and two blocks of Delhi. DDT, BHC carbamate, Endosulfan, etc., are the most common pesticides used in India. But, the vulnerability of groundwater to pesticide and fertilizer pollution is governed by soil texture, pattern of fertilizer and pesticide use, their degradation products and total organic matter in the soil.

Pollution of groundwater due to industrial effluents and municipal waste in water bodies is another major concern in many cities and industrial clusters in India. A 1995 survey undertaken by central pollution control board identified 22 sites in 16 India as critical for groundwater pollution, the primary cause being industrial effluents.

Nitrate Pollution

The problem of nitrate pollution in groundwater becomes severe in the country. The state of Maharashtra alone showed severe problem of nitrate pollution. The studies carried-out by central ground water board in the state revealed that, out of 688 samples, 75% sample as have shown nitrate levels below desirable limits, while 14% samples have nitrate levels above 100 mg/l.

There is insufficient data for making categorical statements on the status of nitrate pollution in India. A recent publication indicated that nitrate in groundwater was in the range 0.1 – 870 mg/l with an average of 65 mg/l 150. Gujarat, Rajasthan and West Bengal were the states with high nitrate content in groundwater. A world bank – assisted study in Karnataka showed that nitrate content in groundwater. A World Bank assisted study in Karnataka showed that nitrate contamination of groundwater was a problem in a majority of districts 51. In the year 2000, it has been reported that more than 33 per cent of water samples in Punjab and Harayana has nitrate levels above the desired limits 52.

Management Options to Reduce Nitrate Leaching from Agriculture

Among the various option available to reduce nitrate leaching are cover crops, using specific management practices, use of nitrification inhibitors or controlled-release fertilizers, site-specific crop management, viable rate application , proper scheduling of irrigation sprinkler or drip irrigation with fertigation, banding and split application, chlorophyll measurement or colour chart based nitrogen application, foliar application, etc. Enhancing fertilizer N use efficiency by ensuring measures such as balanced application of nutrients, proper coordination of N an irrigation management and site- specific and need based nutrients management can substantially control the leaching of nitrate N beyond root zone of crops. There can be two general approaches to minimize environmental pollution arising due to N use in agriculture: one is the optimum use of the ability of crop plants to compete with other processes which lead to the losses of N from soil plant system to the environment and the other is the direct reduction of the rate, duration and extent of loss through the loss processes themselves. Balanced application of N, P and K can significantly reduce the amount of unutilized nitrate in the root zone.

A comparison between organic and conventional farming found that nitrate losses from arable crops averaged 47 and 58 kg N ha-1 for organic and conventional farming system respectively, and concluded that under similar cropping, losses from organic system are slightly smaller than that those from conventional farms following best practice.

Environmental Impact

The changes in environment caused due to agricultural practices varied : degradation of land, 173 million ha (53 per cent) out of 329 million ha of total geographical area, deforestation, loosening of soil structure, depletion of soil nutrients, micronutrient deficiencies water logging, depletion of ground water resources changing pest, disease, weed syndrome, eutrophication of water bodies and methaemoglobinaemia (nutrient accumulation). Apart from this the effect of "Biocides" such as insecticides, herbicides, fungicides and rodenticides remain active long after their targets are destroyed this makes "Biocides" environmentally dangerous. Through biological magnification their concentration in food chain gets magnificed at each tropic level.

India with 18 per cent of world population and 15 per cent of total livestock population depending on 2.2 per cent of world geographical area with the vast population base, food needs and related burden on agriculture are increasing. This necessitated for sustainable agricultural development.

Sustainable Development

At the world summit on sustainable development in the Johannesburg, South Africa (WSSD August 26th September, 2002) the international community was challenged to fulfill the vision of the Rio earth summit" by gearing up to undertake and complete the unfinished business of the sustainable development agenda. WSSD document said that, international community is challenged to alleviate poverty and achieve food security while ensuring the sustainable use of land and other natural resources.

Agriculture. environmental measures offer opportunities for favouring the build-up of soil organic matter, the enhancement of soil bio diversity, the reduction of soil erosion, contamination and compaction. These measures include support to organic farming, conservation tillage, the protection and maintenance of terraces, bio-pesticides, use of integrated crop management,

management of low intensity pasture systems, lowering stock density and the use of certified compost.

Biodiversity

Agricultural biodiversity encompasses the variety and variability of animals, (including agricultural forestry) plants and micro0organisms at genetic species and ecosystem levels necessary to sustain key functions of the agro-ecosystem, its structure and processes for and in support of food production and food security (FAO, 1990).

These locally diverse food production systems are under threat and with them, the accompanying local knowledge, culture and skills of the food producers. With this decline, agricultural biodiversity is disappearing and the scale of loss is extensive and with the disappearance of harvested species, varieties and breeds goes a wide range of un harvested species.

1. More than 90 per cent of crop varieties have disappeared from farmer's fields; half of the breeds of many domestic animals have been lost.
2. In fisheries, all world's 17 main fishing grounds are now being fished at or above their sustainable limits, with many fish population effectively becoming extinct.

Causes of the Losses of Agricultural Biodiversity

There are many causes of this decline which has been accelerating throughout the 20th century in parallel with the demands of an increasing population and greater competition for natural resources. The principal underlying causes include:

- The rapid expansion of industrial and green revolution agriculture, intensive livestock production, industrial fisheries and aquaculture (some production systems using genetically modified varieties and breeds) that cultivate relatively few crop varieties in monoculture rear a number of domestic animal breeds or fish for or cultivate few aquatic species.
- Globalization of the food system and marketing and the extension of industrial patenting and other intellectual and rearing of fewer varieties and breeds for a more uniform less divers but more competitive global market.
- Marginalization of small-scale diverse food production systems that conserve farmers varieties of crops and breeds of domestic animals which form the genetic pool for food and agriculture in the future.

- Reduced integration of livestock in arable production which reduces the diversity of uses for which livestock are needed.
- Reduced use of 'nurture' fisheries techniques that conserve and develop aquatic bio-diversity.

Variety replacement is the main cause of losses. The replacement of local varieties or landraces by improved and / or exotic varieties and species is reported to be the major cause of genetic erosion around the world it is also cited as the major cause of genetic.

Agricultural biodiversity provided many ecological services within agro-ecosystems such as nutrient cycling pest and diseases regulation and pollination. While bio-diversity "richness" differs according to climate, terrain farming practices and so on. Farm based on multiple crops and livestock with natural pasture are richer in bio-diversity than mono-cultural farms. However, most systems by seeking to maximize the yield of limited number of plant and animal species, inevitably weaken and reduce competition from unwanted species.

The major changes in agriculture have upset its equilibrium with biodiversity. There are the intensification of production, on the one hand and the under – utilization of land, on the other. The specialization, concentration and intensification of agricultural production that have occurred during the last decades are widely recognized as potentially threatening bio-diversity conservation. Many species have a direct interdependence with agriculture (i.e. many bird species nest and feed on farmland). However, it is difficult to isolate the effects of changes in land use from others such as urbanization and the progression of infrastructure, which also occur in rural areas. However, sound agricultural management practices can have a substantial positive impact on the conservation of wild flora and fauna, as well as social–economic situation of rural areas.

Ecological Agriculture

A gradual shift from chemical to ecological farming may be a viable option to make agriculture sustainable. Immediately, the farmers could build up a sufficient organic base to fertilize the fields and improve soil productivity. The cost of cultivation can be brought down substantially since many farm-growing inputs can be integrated efficiently into the farming system. Ecological farming is a highly knowledge – intensive, labour – oriented and complex system integrating several organic recycling processes.

Eco-friendly microbes, particularly, bio-fertilisers, such as Rhizobium and Azotobacter are being used by farmers for a number of crops. Sprinkler and drip irrigation will ensure optimal and timely irrigation along with effective land preparation. Questioning the systems of mono – culture is the demand for eco-generation by way of mixed cropping patterns. Environmental-friendly pesticides such as neem derivatives have become important components of integrated pest management strategies. The problem of seepage of canals can be tackled best by lining and distributions with concrete. Many experts have recommended stall – feeding to over come problems of overgrazing.

Ecological farming is economically rewarding, intellectually stimulating and energy-saving system. It is also a net foreign exchange saving system since it reduces the import of fertilizers. Sustainable farm practice with the help of integrated pest management, integrated nutrient management and integrated weed management is the call of the hour. At Dharwad crop residue in corporation with *Phenorocheate chrysosprium*, FYM, Vermicompost in on integrated nutrient system produced similar yield in cotton and soybean as comparable to in organic fertilizers, improvement in soil physical, chemical and biological properties was observed (Babalad, 1999)

Organic Farming

Over the years the traditional and the indigenous knowledge on agriculture was faded away. The conversion to organic farming in some corner is seen and advocated as coming back to traditional agriculture in which by default they were not using hybrids, chemical fertilizers and pesticides, because they were not available before 1950s. The concept of organic farming is more intrinsic and scientific than the conventional agriculture. Traditional agriculture does not completely fit into organic agriculture as it did not involve maneuvering of soil, plant, environment and people to achieve sustainable production.

Organic farming is not just exclusion of chemical fertilizers and pesticides. It is basically a farm activity which are biological systems and not mechanical ones. The design and application of organic agriculture recognize this essential biological character, works with rather than against natural ecosystems and often in ways that create job as well bring sustainability. Organic farming is exclusively a certification programme which consists of standards, production and processing technology package, certification and regulatory mechanism. Certification in organic agriculture generally can be done independently or third party. The system includes farm inspection and audit trails (keeping and checking records).

The philosophy behind the concept of organic farming is to feed the soil, rather than crop and it is a means of giving back to nature what has been taken from it. Organic farming does not refer to only input management, this treat the farm as an organism and holism. The concept of organism means, in which all the components – the soil, organic manure, micro-organisms, insects, plants, animal and human interactions to create coherent self regulating and stable whole.

Ecological Services

As it excludes the use of chemical fertilizers and pesticides from the perspective of eco-service management has many advantages over conventional agriculture. We can no longer afford to ignore the need to manage ecological services more actively focusing on farming, as indicated in the "Millennium ecosystem assessment report". A recent study by Pimental *et al.*, (2005) at Rodale Institute over 22 years showed greater advantage from animal based organic system than conventional system in maintaining health of agro-ecosystem.

Research Initiatives

Several Institutes initiated research programmes on organic farming. Under the aegis of ICAR, Project Directorate for cropping system has initiated Network' Project on Organic farming with 12 co-operating centres covering 12 states of the country. The programme was initiated to study the productivity, profitability, suitability, quality and input use efficiency in different crops and cropping system under organic production and compared with conventional systems across the country. The efficient crop and soil management practices in different crops has been taken up At several locations, yield under organic system in the crops mainly groundnut, soybean, chickpea, basumati rice, maize, cotton, vegetable spices were comparable with integrated practice with greater sustainable and environmental benefits (Anonymous, 2006).

The long-term experiments at ICRISAT, Patancheru India, since 1999 low cost input system (organic) showed similar yield that of conventional agriculture. Low cost input system which depends on scientific and traditional knowledge of farmers were crop residues, bio-fertilizers, compost, green manures, bacterial inoculants, botanical extracts as nutrients and bio-pesticides helped in maintaining soil fertility and crop yield (Rupela et al., 2004).

Number of studies showed that under drought conditions crops in organic system produced significantly higher yields than conventional agricultural crops often out-yielded conventional crops by 7-90 per cent.

National Importance

The today's, mechanization is not relevant to small holding. In India at least 74 per cent farmers own <2 ha. These small holding can adopt the organic principals easily and on farm management of inputs and labour more efficiently. Drylands occupy 65 per cent of the total cultivable land. In traditional rainfed agriculture the average national productivity is 1.0 t/ha which can be enhanced to 2.0t/ha and organic agriculture has shown potential to increase yields. Under the present fertilizers crises organic agriculture has advantage as India has great potential to substitute chemical fertilizers through organic fertilizers. The existence of traditional knowledge and farming systems has strong linkage between agriculture, livestock and other which can provide livelihood security to farm families and replenish with the soil fertility. Varied agro climatic conditions and longer growing period are as prevailing in India has an advantage to meet the off season growing demand of export market and also the domestic market. The market for organic food is also growing as the awareness about quality food among the consumers is increasing. The organic sector is growing at 25 to 30 per cent annually will create employment opportunities for production, processing and value chain at rural area and also marketing in cities.

Integrated Pest Management

In agro- ecosystems, ecological interactions between the environment, plants, herbivores (diseases, insects and rats) and natural enemies of herbivores take place. The herbivores are balanced by their natural enemies. The inter dependencies need to be understood well to know when pest control is really needed to prevent economical losses and maximize profits.

Recently in intensive agriculture, use of chemical is considered as most common method of pest management. Because of its indiscriminate use, many insects have developed resistance to some of the commonly used insecticides. *Helicoverpa* is one of the good example which has developed resistance. Apart from this, taking into account of the hazards due to excess use of insecticides leading to pollution of soil, water and air which has direct relation with the human health.

Integrated pest and disease management practices have made it possible to reduce and sometimes even eliminate pesticide sprays, substituting eco-friendly control practices and organic pesticides such as the nuclear poly hydrosis virus. By monitoring pest populations and spraying only on a need basis, farmers save money and reduce health risks. Use of more than one components of pest management such as use of botanical pesticides, bio-agents like practices, predators and viruses causing diseases in insects, modification of cultural practices, use of trap crops, use of pheromone traps and light traps, use of pest resistant crop verities, mechanical methods like collection and destruction of insect pests, etc., keeping the use of chemical insecticides as a last resort and at minimum level which is more effective in pest management.

References

Anonymous, 2006. Annual Report, Network Project on Organic Farming PDCSR, Modipuram (ICAR), University of Agricultural Sciences, Dharwad.

Babalad, H.B., 1999. Integrated nutrient management in soybean based on cropping system. Ph.D thesis submitted to the University of Agricultural Sciences, Dharwad.

ORCD (2001), environmental indicators for agriculture volume 3 : methods and results ORCD Paris.

Pimental. D., Hepperly, P., Hanson, J., Douds, D and Seidel R., 2005. Environmental synergistic and economic comparison of organic and conventional farming system. *Bio. Sci.* 55 (7): 573-582.

Ramanathan, K.M., 2006. Organic farming for sustainability. *J. Indian Sol. Soil Science.* 54 (4) : 418 – 425.

Rupela, O. P., CLL Gowda, S. P. Wani and G. V, Ranga Rao,2004. Lessons from non chemical input treatments based on scientific and traditional knowledge in a long – term experiment. Proc. International conference on agricultural Heritage of Asia, 6-8 Dec, 2004. *Asian Agri-History Foundation, Secandrabad, Andra Pradesh.*

Tripathy, I.G and K. K. Tripathy, 2009, Agriculture and Environment—Case for sustainability. Business line, Wednesday, November 10, 2004.

3

Sustainable Agriculture and India—Dimensions and Directions

V. Basil Hans

Introduction

Agriculture in India is in a peculiar situation of growth with vulnerability. While the contribution of agriculture to Indian economy's growth and development has not been meagre, agricultural and rural backwardness along with a growing trend of human poverty reflected in financial crisis and the trauma of farmers' suicides are causes of concerns. Any agenda for modernisation of Indian agriculture should have the objectives of ensuring equity and sustainability. Sustainability on the one hand can not only make any effort of raising agricultural productivity and farm incomes possible and useful, but can also ensure a balanced growth of non-farm activities in the economy.

Sustainability must viewed in the backdrop of a paradoxical situation in India: On the one hand there are record food grain stocks standing at an all-time high (62 million tonnes against an annual requirement of around 20 million tonnes for ensuring food security), and on the other hand, over 200 million of India's population is underfed, and millions are undernourished. There is a widening gap between the elites forming a tiny minority and the mass of common people suffering from multiple deprivations (Rao, 2005). The challenge is to bridge this gap. In a scenario of shrinking land and depleting water resources, the challenge of the new millennium is to increase biological yields to feed the ever-growing population without destroying the ecological foundation. It is thus important not to package this challenge as a demand or imposition of society on farmers, for which farmers would bear the cost, but as a necessity and methodology to also sustain their welfare and income (http://www.w.envfor.nic.in/divisions). Another challenge is to save Indian agriculture from becoming non-competitive under globalisation. The third challenge is to keep Indian agriculture always prepared to face any consequences of industrial recession.

The bright side of the picture is that India has the potential to meet these challenges. Already some policy affirmative and actions are evident. Sustainable agriculture is not only an objective of the government and a creative arena of agricultural scientists, but also a desire of most Indian farmers.

Sustainability—Conceptualization

'Sustainability' is one of the buzz-words of our times. It has been attached to any number of social and economic pursuits: we want sustainable economic growth, sustainable development, sustainable forestry, a sustainable population, sustainable cities, and so on. Increasingly, we also talk about the need for "sustainable agriculture". But just because we use the word in abundance doesn't mean we're any closer to achieving it in practice, or even in understanding what it means. In agriculture, we need to define it in terms that farmers can understand and to develop ways in which they can measure it. Above all, perhaps, we need to ensure that the concept is useful to farmers (indiancommodity.com, n.d.).

The idea of sustainable agriculture has been around a long time. Since the very first crop was sown and animal was penned, farmers have tried to ensure that their land produces a similar or increasing yield of products year after back-breaking year; recent attempts to popularise the concept build on this tradition.

Sustainable Agriculture is a special kind of agriculture along with farming technique which makes maximum utilisation of the environment without causing any form of harm to it. The outputs from this Sustainable Agriculture are devoid of any types of inorganic chemicals such as pesticides and insecticides. This farming technique uses the organic way of farming. All these factors produce the products in a more environment friendly manner and are thus healthier for the consumers to consume (Economy Watch, n.d.). One modern definition is as follows:

Sustainable agriculture is the use of farming systems and practices which maintain or enhance:

- the economic viability of agricultural production;
- the natural resource base; and
- other ecosystems which are influenced by agricultural activities.

This definition can be supplemented by some fundamental principles of sustainable agriculture:

- that farm productivity is enhanced over the long term;
- that adverse impacts on the natural resource base and associated ecosystems are ameliorated, minimised or avoided;
- that residues resulting from the use of chemicals in agriculture are minimised;
- that net social benefit (in both monetary and non-monetary terms) from agriculture is maximised; and
- that farming systems are sufficiently flexible to manage risks associated with the vagaries of climate and markets (ibid).

The other characteristic features of sustainable agriculture are:

- The treatment towards the animals in such kind of farming are more humane because they are given enough space to graze and are not bred like factory products.
- The workers who are associated with the sustainable agriculture receive more humane behaviour from the employers and it has been observed that the wages offered to these farm workers are fairer than the ones working in the so called industrial farms.
- The one of the most important feature of sustainable agriculture is its extended support towards the small farming community farmers. It is essentially being emphasized in this type of farming that the organic food items that have been produced by the small farmers of the rural areas are bought and marketed to the consumers through this Sustainable Agriculture route (ibid).

Thus, sustainable agriculture is an integral part of sustainable development which in its broadest sense is concerned with resource availability/accessibility/utilisation/conservation so as to maintain a health relationship between man and nature now and in the future. In the context of inter-generational and international equity considerations and environmental concerns, the role of agriculture has become not only to increase agricultural production so as to optimally utilise national resources but also ensure stabilisation of the growth and allowing the different participant sub-sector opportunities and accessibility to development. It is efficient management of agronomy for the sake of balanced development. It is a humane approach to agriculture to ensure a better quality of life for the people and other resources (Hans, 2007a; Hans, 2007b; Giriappa, 1999).

Sustainability in India: Practical Issues

Any programme for sustainable agriculture in India should be preceded by a SWOT analysis of the sector, particularly in the globalisation era. As Nehru once said, agriculture is India's greatest living industry. Transformation is still alive in this industry (Hans, 2006). It has a record of scripting its own success despite odds like falling water levels, expensive credit, a distorted market, laws that stifle private investment, controlled prices, poor infrastructure, produce that does not meet international standards, inappropriate research and tax evasion by unorganised sector leading to the lack of a level playing field (Dwivedi, 2009). From a largely traditional farming in the early 60s, to high-tech agriculture and agribusiness, from cooperative farming to contract farming, from macro-financing to micro-financing, the sector has seen many changes. With changes have come challenges too.

As mentioned earlier, Indian agriculture is – like the Indian economy itself—in a paradoxical situation, viz., poverty in the midst of plenty. The country is endowed with 183 million hectares of cultivable land, 115.6 million farming families, 400 million of annual precipitation, and a conducive agro climatic advantages for the cultivation of a number of crops. Nearly two-thirds of our population work in the agriculture and feed over 1000 million people in the country. Yet majority of them are under the clutches of poverty, debt and hunger. Several thousands of farmers in Vidharbha (Maharashtra) and other parts of India have committed suicide in the recent years.[1] Agriculturists in general and the small and marginal farmers in general have been the worst sufferers from the onslaught of globalisation. With more than 40 percent of agricultural lending even today coming form the non-institutional sources charging anywhere between 30-40 percent interest per annum, the farmers are in an immiserising situation. The NSSO Report released in 2005 states that 1 in 2 farm households are in debt and only 10 per cent of the debt was incurred for non-production purposes. The National Crime Records Bureau reports that between 1997-2005 1,56,562 farmers committed suicide. Nearly 60 of them took place in the 4 progressive states, viz.,

---

1. Despite a multi-crore relief package announced by the Centre four months ago, the graph of farmer suicides in Vidarbha region continued to rise alarmingly. Sustainable agriculture has shown a way out of the problem here. Yuva, a Nagpur-based NGO, says that using an Integrated Natural Sustainable Agricultural Process (Insap) has helped them successfully combat farmers' suicide in over 500 villages of Vidarbha. This included alternative methods of soil and water amangemnt, as well as marketing. Input cost got reduced by 50 per cent! About four lakh farmers in five Vidarbha districts Amravati, Akola, Buldana, Washim and Wardha are using eco-friendly methods. "In these villages, no farmer has committed suicide," says Palash Ghoshal of Yuva (Times of India, 13 Nov 2006).

Maharashtra, Andhra Pradesh, Karnataka and Madhya Pradesh. More than 20 per cent of suicides have taken place in Karnataka (Mahalingam, 2007; Hans 2008a).

The establishment of World Trade Organisation (WTO) in 1995 – to make the world trade rule-based, transparent and free – had major implications for India and its agriculture. Initially the reforms did result in improvement in terms of trade for agriculture thanks to decontrol of fertilisers and a substantial hike in minimum support prices given by the government. Indian agriculture witnessed a small acceleration in the growth rate. The country had successfully overcome the food crisis. But there were other crises to follow. The growth rate in agriculture and allied sectors (at 1993-94 prices) that had risen from 3.3 per cent in 1951-61 to 3.9 per cent in 1981-91 (after a low of 1.7 per cent in 1971-81), stood at 5.8 per cent in 1992-93. But in 1993-94 it fell to 4.1 per cent. In 2006-07 it was 2.7 per cent and for the entire period 2001-07 it was only 2.2 per cent. Investment in agriculture as percentage of GDP fell from 1.92 per cent in 1990-91 to 1.7 per cent in 2004-05. Capital formation in agriculture which was 1.9 per cent of GDP in early 90s declined to 1.3 per cent in 2000-01. The share of agricultural exports in total exports declined from 19.4 per cent in 1990-91 to13.5 per cent in 2000-01. Come to think of it, the share was above 44 per cent during the 60s. Unemployment in the agricultural sector has been a bane during the reform period as agriculture was not profitable due to the fall in the price of farm products. When the number of people employed in the primary sector and the area under cultivation decreased it caused a decline in rural employment. According to the National Sample Survey, the annual rate of growth of the employment in the rural areas was 2.07 per cent in 1987-1984, while it declined to a mere 0.66 per cent in 1993-2000, which corresponds to the period of liberalisation (Hans, 2008a). The trend in rural-urban migration goes on unabated what with land being increasingly diverted to non-agricultural uses under the banner of SEZs.

The performance of the agricultural sector in the post globalisation era has been so much unsatisfactory that some feel that this clearly shows the beginning of the end of the end in the growth of agricultural sector. Yet this is not to degrade the sector or to entertain the deficient growth rate. On the contrary it is to enlighten the necessity of boosting its performance and in essence to have second Green Revolution in the country (Vinayagamoorthy and Thirupathi, 2007). This is also a pointer to sustainable agriculture as a goal.

A decline in the share of agriculture in the national income form over 50 per cent during the 50s to less than 20 per cent today may be a sign of structural transformation but the question today is whether Indian agriculture will be able to meet the new demands placed upon it by Liberalisation, Privatisation and Globalisation (LPG). By the year 2030, India will have to produce 60 percent more rice with much fewer resources. Clearly, there will be a major challenge for scientists and policy makers to meet the increased food demand. With diminishing land and other resources available and need for additional food in the coming decades, a deficit in agricultural production already predicted India has to play a major role not only in maintaining its own self-sufficiency in food production but also to meet the additional requirements of its neighbours. While globalisation and liberalisation have increased the commercial proposition of this sector, it is facing a number of challenges with regards to technology, infrastructure, price and competition. There is declaration of output and acceleration of vulnerability of farmers to poverty. Modernisation without moderation has led to a crisis situation in Indian agriculture (Hans, 2008a). This manifold crisis in Indian agriculture is a reflection of low productivity on the one hand and withering livelihood security on the other. The situation must also be juxtaposed with the challenges from climate change and global meltdown.

According to Dr. Shiraj A Wajih, President of Gorakhpur Environmental Action Group and member of State Agriculture Committee of Uttar Pradesh, "Increasing cost of agricultural production material, climatic changes, chemical farming, poor management of agriculture system, lack of government interest and low market value of agriculture produce– all these factors tend to worsen the situation for small and marginal farmers....Sustainable agriculture literacy can alleviate the problem of agriculture in India. We need to educate the farmers. Farmer schools can be a good measure of sustainable agriculture literacy. This school should be run by the community members and local farmers can be the best teachers and trainers of such schools" (Dwivedi, 2009).[2] They ebale the farmers and raise their socio-economic status.[3] The need of the hour is to make agriculture sustainable in a wholesome way.

2. As Dr. Shiraj says, farmer schools can be a good measure of sustainable agriculture literacy. Such a school should be run by the community members and local farmers can be the best teachers and trainers. Gorakhpur Environmental Action Group is currently running 12 farmer schools in two districts of Uttar Pradesh. All these schools have proved to be a unique platform to resolve the problems of agricultural production faced by small and marginal farmers.

3. "Who said agriculture is a business of loss. I am saving more than Rs 80,000/- (about USD 1,750) every year and providing bread & butter to my 11 family members who are dependent on me," said Prabhavati Devi, who lives in Sardar Nagar block of Gorakhpur district. She has one and half acres land in which she is producing more than 86 types of crops annually. Prabhavati Devi is doing organic farming, which is a very reliable method of sustainable agriculture resulting in high production at low cost. According to her, '12 years back I too was doing chemical farming and had no idea about this model of sustainable agriculture which involves an efficient management of time and land available for farming. (However, ustainable agriculture should not be confused with organic farming as both are very different form each other.) See Dwivedi, Amit, 'Sustainable agriculture can be the back-bone of Indian economy' *Northern Voices Online*, Nov 26th, 2008.

From the quantitative side the major challenge of the 11th Five year Plan is to raise the growth rate of agricultural GDP to at least 4 per cent to ensure a 7 plus growth rate of the economy as a whole and to ensure a broader spread of benefits (Murugan and Anbumani, 2007). While this calls for a host of modern practices our farmers must adopt modern practices of farming with a pinch of salt. One cannot simply sail on the credentials of green revolution. Sustainable agriculture should be the objective (Hans, 2008a). The sustainable agriculture strategy has to be multi-dimensional. It is must be eco-friendly and pro-farmer with a proper balance between the economic, political, social, cultural and emotional aspects of farmers and farming. Saving farmers to save farming must be the agenda of action.

Structure and Strategies

Some tend to confuse sustainable agriculture with organic farming. But both are very different form each other. Sustainable agriculture means not only the withdrawal of synthetic chemicals, hybrid-genetically modified seeds and heavy agricultural implements (as in organic farming), it also tries to simulate the conditions found in nature. Sustainable agriculture involves multiculture, intercropping, use of farmyard manure and remnants, mulching and application of integrated pest management. If this is followed then there is no reason why agriculture cannot be an economically viable activity in addition to being environmentally sustainable (Dwivedi, 2008). That means sustainable agriculture is not only ecological", and "organic", or "bio-dynamic" but also "low input (cost)" (alchemistjk, 2007).[4]

Sustainability underpins the need for integrated use of inputs, a thing that is not entirely new in India. For instance, we had a system of integrated pest management even in the past. Even while chemical fertiliser becomes inevitable, the average farmer is to use it in the recommended ratio based on soil-water-crop management techniques. Educated farmer also identifies pests and diseases better and has a proper pest management schedule. He incorporates natural processes such as nutrient recycling, nitrogen fixation and pest-predator relationships into agricultural production processes. Biological control of pests and diseases and use of bio-pesticides is an integral part of pest management system (Raju, 2007; Krishnakumar and Purushothama, 2005). It is therefore a vital feature of sustainable farming. In fact to operationalise sustainable agricultural development

4. However, it started as started the modern organic farming system in the 1930s by Sir Albert Howard in England emphasizing the use of compost instead of chemical fertilisers (alchemistjk, 2007).

three steps are very essential: (i) appropriate system, (ii) appropriate technologies, and (iii) institutional arrangement. Each farmer and each region must have an identified agro-eco-system ideal for the goal of sustainable agricultural development.[5] Here efficiency criteria include physical productivity and cost effectiveness developed into alternative models based on the information and empirical evidences from research. In one such study Nadkarni points out that ecological farming outperforms two other alternatives, viz., the green revolution model sponsored by the state under well assured irrigation and the traditional farming pursued by the majority of dryland farmers. That was mainly because green revolution model relies heavily on petro-based external inputs the prices of which are rising and also they are having much problem from externalities whereas eco-farming depends mostly on internal inputs (Prakash, 2005). To keep up the momemtum of agricultural growth a careful evaluation of inputs like seeds, fertilisers and irrigation sources (e.g. micro-irrigation technology) are of paramount importance (Kiran et.al, 2009).

Another perspective of sustainable agricultural development that must be the focal point in the coming years is linked to the cliché slogan "think globally and act locally". Macro growth without micro stability is harmful. Government governance and global governance must be harmoniously blended to achieve good governance (Hans, 2002). Local participation in agro-planning reduces time and money of centralised and trickle down strategies. As far as global measures are concerned the issues of subsidy and liberal imports must be negotiated with all earnest so that agriculture is not discriminated and barriers to agricultural growth are removed.

At the local level what works better in the realm of credit is micro-finance. Microfinance services are financial services that poor people desire and demand (i.e. they are willing to pay) for. The term also refers to the practice of sustainably delivering those services (Hans, n.d.). Both on-farm and off-farm facilities need to be synchronised in usage. For instance, better marketing and infrastructural facilities will add to the benefits of on-farm efficiency. Visible improvement has come about through adoption of management practices through on-farm and off-farm operations in this sector. Agri-business has evolved out of the new input-

5. Wherever feasible due to more common features – physical and otherwise – of farming, individuals may form economic units of farming and use resources commonly like in the case of a village watershed. They can be joining together of private sector and the third sector. Even governmental and non-governmental agencies may join hands in the larger interests of healthy farming. A public-private partnership can be made functional. Fine balance of local resources and local participation – adoption and adaptation of traditional and modern practices becomes possible.

output matrix. Agriprenuers have come to realise the importance of quality changes and value addition in agriculture. The challenges of the rural economy can be balanced provided there is better management in the economy. Right type of operational and managerial skills and entrepreneurial expertise peppered with timely measures by the government would help fulfil the growing needs of agri-business and sustainable agriculture. The government has already articulated this in the National Agricultural policy 2000. The New Agricultural Policy has both quantitative and qualitative targets for Indian agriculture. The time has therefore come to kick start a revolution of a third kind. This revolution will have to be a modified one. It has to encompass all the resources, regions and activities of the rural economy. Its economic benefits should reach the poor through rural value-added enterprises and partnerships with the private. It cannot be exclusively input management or output management but a blend of both. The country needs a healthy combination of gene revolution and *gram* (village) revolution (Hans, 2008b)

Thus what is pragmatic and productive for Indian agriculture today is "integrated farming system (IFS) approach". The IFS is a micro, innovative approach to promote integrated use of land and other resources (including technologies) with managerial expertise by the farmers for producing and marketing the products to get sustained net income (Nagarajan, 2005).

Conclusion

To conserve resources and to raise productivity in agriculture and ensure a better livelihood to the farmers in India, there can be no better option than sustainable agriculture. While government should make it the central element of agricultural planning, farmers should take effective measures to make it a plank for higher trajectories of growth of productivity and incomes. They must do this also to reduce the degree of dependence on others as well as their vulnerability to the forces of nature and market. They must also be prepared to change further and for the better as they have indicated in the past.

References

Alchemistjk (2007), 'Low Input Sustainable Horticulture', *Agriculture Information Online Business Community*, December 24 < http://www.agricultureinformation.com/forums/organic-farming/17784-sustainable-organic-agriculture.html> (accessed March 28, 2009).

Dwivedi, Amit (2008), 'Sustainable agriculture can be the back-bone of Indian economy' *Northern Voices Online*, < http://nvonews.com/2008/11/26/sustainable-agriculture-can-be-the-back-bone-of-indian-economy/ > (accessed March 27, 2009).

——————(2009), 'Sustainable agriculture literacy mission, welfare of Indian farmers', *Northern*

Voices Online, <http://nvonews.com/2009/01/20/sustainable-agriculture-literacy-mission-welfare-of-indian-farmers/> (accessed March 27, 2009).

Economy Watch (n.d.), 'Sustainable Agriculture'<http://www.economywatch.com/agriculture/sustainable/> (accessed 21/3/2009).

Ghosh, Avijit (2006), 'Sustainable agriculture can check farmer suicides', *The Times of India*, November 13 <http://timesofindia.indiatimes.com/articleshow/msid-421546,curpg-1.cms > (accessed March 27, 2009).

Giriappa, S. (1999), *Prospects of Hi-tech Agriculture*, New Delhi: Mohit Publications.

Hans, V. Basil (2002), *Money, Banking and Public Finance*, Mangalore: Mangala Publications.

Hans, V. Basil (2006), "Towards a Vibrant Indian Agriculture", *Kisan World*, 33(06), pp. 18-19.

Hans, V. Basil (2008a), "Indian Agriculture in the Globalisation Era – Position and Prospects", Paper presented at the UGC Sponsored State Level Seminar On "Impact of the Globalised Economy of India on Agriculture" organized by the Department of Economics, Milagres College, Kallianpur, D.K., Karnataka State on January 24-25, 2009.

Hans, V. Basil (2008b), 'Agribusiness and Rural Management in India – Issues and Challenges', < http://ssrn.com/abstract=1271622>

Hans, V. Basil (n.d.), Microfinance – From Financial Intermediation to Social Intermediation. In Rasure, K.A., (ed.) *Rural Credit in the Era of Globalisation* , Delhi: Abhijeet Publications, forthcoming.

Hans, V. Basil and Jayasheela (2007a), Environmental Management and Sustainable Development in India: Issues and Challenges. In Rasure, K.A. (ed.), *Environment and Sustainable Development*, New Delhi: Serials Publications, pp. 263-74.

————————(2007b), "Environmental Management and Sustainable Development – Implications for Indian Agriculture", *Financing Agriculture* 39(4), pp. 20-25.

Indian Commodity (n.d.), 'Developing Indicators for Sustainable Agriculture' <http://www.indiancommodity.com/ResSustainreport.htm> accessed 21/3/2009.

Kiran, V. Basil Hans and Jayasheela (2009), Indian Agriculture: Crisis and Challenges under Globalisation', *Social Action*, 59(1), pp. 106-15.

Krishnakumar Suchitra and Purushothama, M.K. (2005), Sustainable Agriculture: Concept and Strategies. In In Ramesh, R.S. et.al (ed.), *Sustainable Agricultural Development in India*, Bangalore: A Southern Economist Publication.

Mahalingam, N. (2007), "Agriculture in India: Paradoxes Galore', *Kisan World*, 34(05), pp. 3-7.

Murugan, C. Siva and Anbumani V. (2007), "An Approach to the 11th Five-Year Plan Agriculture to get top priority", *Kisan World*, 34 (02), pp. 36-39.

Nagarajan, G.N. (2005), Integrated Farming System Approach for Sustainable Agriculture. In In Ramesh, R.S. et.al (ed.), *Sustainable Agricultural Development in India*, Bangalore: A Southern Economist Publication.

Prakash, T.N. (2005), Sustainable Agricultural Development Strategies for India. In Ramesh, R.S. et.al (ed.), *Sustainable Agricultural Development in India*, Bangalore: A Southern Economist Publication.

Raju, S.S. (2007), "A Frame-Work of Sustainable Agricultural Development in the Context of Globalisation", *Journal of Global Economy*, 3(1), pp. 33-36.

Rao, V.M. (2005), Sustainable Agricultural Development. In Ramesh, R.S. et.al (ed.), *Sustainable Agricultural Development in India*, Bangalore: A Southern Economist Publication.

Vinayagamoorthy A., and Thirupathi T. (2007), "Globalisation and Agricultural Sector in India", *Kisan World*, 34(06), pp. 37-38.

4

Characterization and Utilization of Natural Resources for Planning at Micro Level for Sustained Productivity in Rainfed Agro-Eco System

M.B. Guled, M.S. Shirahatti,
S.R. Mulla and H. Venkatesh

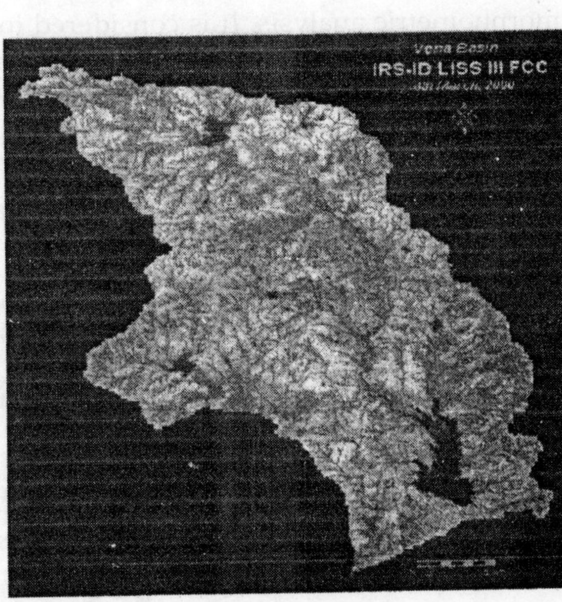

It is realized that the land, water and plant resources need to be characterized, developed, used and managed efficiently in an integrated and comprehensive manner for sustained crop productivity in general for other areas and for rainfed agro-eco systems in particular.

For characterization of soil resources, the synoptic coverage and high resolution of remotely sensed satellite image (Standard FCC imagery of IRS-ID, LISS-III) and topographic data analysis procedures are required. If this procedure is not available, alternatively, the available soft copy of the soil map and drainage map at micro watershed level of the state may be used for preparing the thematic map on soil resource, which will be of broader category. For accurate delineation of different soil units contained in the soil map, detailed soil survey is essentially required. It is expected that, the above cited evaluation will be highly helpful for better land use planning and management at basin/watershed level, and this will also help in the areas of ground water exploration, soil and water conservation and watershed management.

Characterization of rain water for a given locality also assumes greater significance. To achieve this it is required to establish the information on annual rainfall trend over years, return of low/high rainfall epochs, secular changes in inter seasonal and intra seasonal Then variability patterns have to be classified.

A critical analysis and understanding of soil resource and rainwater resource will help in developing soil and water conservation practices, soil quality units, soil suitability criteria for different crops. Crop management practices and contingency crop planning for a given locality. This information can be suitably down scaled even to a panchyat and to be cadastral level of a panchyat.

Characterization of Soil Resource

Method 1: Morphometric Analysis

Remote sensing and GIS techniques are effective tools for conducting morphometric analysis. It is considered to be the most satisfactory method as it enables us to understand the relationships among the different aspects of drainage pattern of the basin/watershed. The morphometric analysis with the satellite data in conjunction with the lithological, geomorphological and land use/land cover properties improve the delineation capabilities it will also help in upgradation of some of the drainage channels, which are altered by the natural forces or human induced activities.

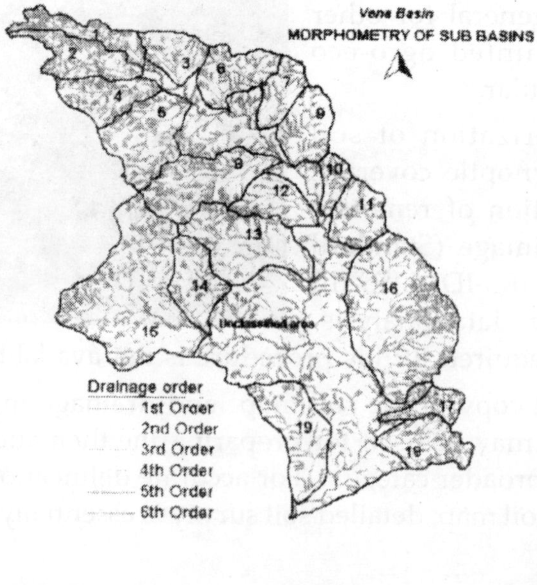

The basin/watershed area will be divided into sub basins/micro-watersheds based on the water divide concept to derive various morphometric parameters.

In the present study the area has been divided into 19 sub basins.

The following morphometric parameters at sub basin level will have to be collected.

- Area of each sub basin (Km^2)
- Perimeter (Km)
- Basin length (Km)
- Drainage order (N_1,N_2, N_3........etc)
- Total number of drainage lines
- Cumulative length (Km) of drainage line
- Log of cumulative length
- Stream frequency (No/sq km)
- Bifurcation ratio
- Drainage density (Km/Sq Km)
- Texture ratio
- Form factor
- Circulation ratio
- Elongation ratio

Making use of the above parameters, the following relationships are established.

- Total number of streams and cumulative length
- Basin area Vs cumulative length of streams
- Cumulative length of Streams Vs Drainage density
- Bifurcation ratio Vs Elongation ratio

The study of the relations as mentioned above result in generating the information on geomorphology, soil depth, soil drainage, Available water holding capacity and soil erosion or different categories of eroded lands.

Slope

The slope analysis reveals that the north-western, western and south-eastern parts of the study area are having moderate (5-10%), strong (10-15%) and moderately steep slopes (16-30%) gradients with an area of about 9.63 per cent,

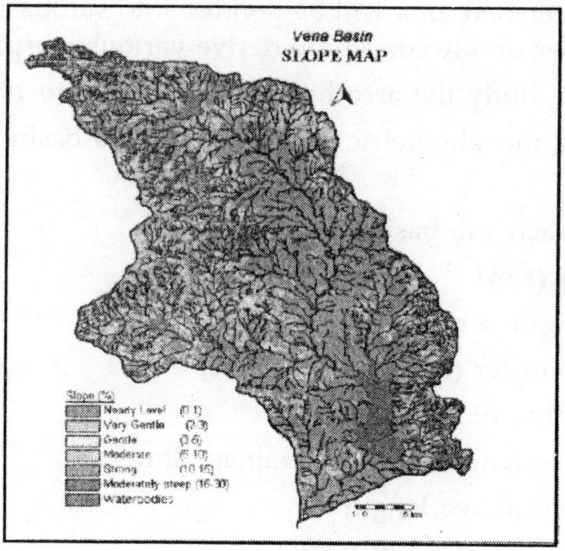

4.89 per cent and 8.05 per cent respectively. This area is mainly covered with dissected ridges, plateau spurs and escarpments. The foot slopes of the dissected ridges and upland areas are associated with gentle (3-5%) to very gentle slopes (2-3%) slopes and account for 7.53 per cent and 14.56 per cent respectively. The majority of the area is under nearly level to level slopes (0-1%) and accounts for nearly 52.08 per cent of the study area.

Geomorphology

In the present study the geomorphological analysis reveals that dissected ridges, isolated mounds, linear ridges, escarpments and plateau spurs were identified and mapped under denudational landforms. Plateau plains (upper sector), middle sector, lower sector (pediplains), narrow valley fills and main valley floors were analysed and mapped under depositional land forms. The gneiss of denudational landforms are influenced by the underlying lithology, slope and morphometry (Subramanyan, 1981). These landforms are associated with high drainage density, bifurcation ratio and high cumulative length of 1st, 2nd and 3rd order streams. The depositional landforms are formed by the influence of permeable lithology, moderate to nearly level plains, medium to low drainage density (<2.0), low cumulative length of streams and are in association with 4th and 5th order streams. It reveals that the type of morphometry, underlying lithology and slope factors have greatly influence the gneiss of landforms.

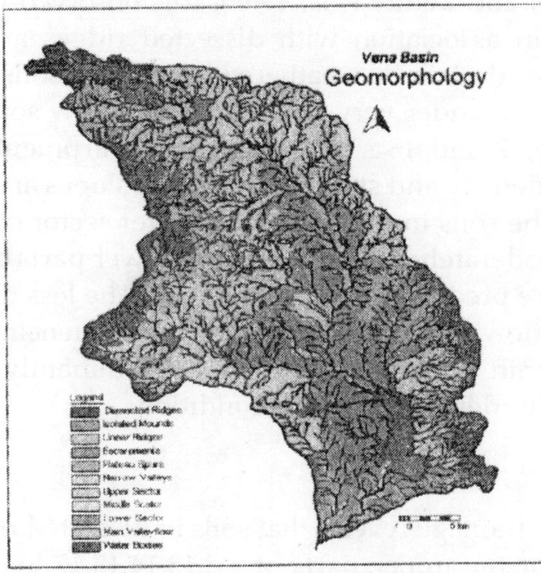

Soil Depth

The relationship among morphometry, geomorphology and soils determine the pedological processes, resulting the evolution of present landscape. It is also well known fact that different geomorphic features in conjunction with type of parent materials and local drainage pattern influence greatly the type of soils formed in topographic sequence under specific geo-pedological environmental conditions

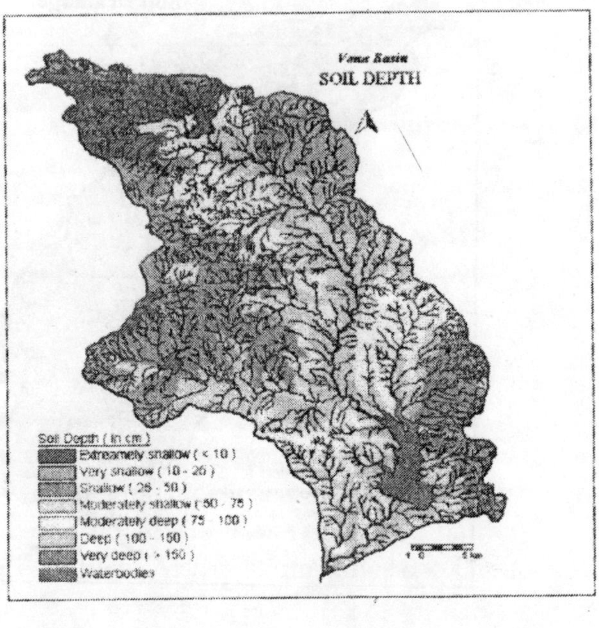

(Das and Roy, 1979, and Reddy *et al*, 1999). It is observed that, soils in the 1,2,3 and 4 sub basins in association with dissected ridges and plateau spurs are extremely shallow underlain by weathered basalt. The soils in the upper parts of 7 and 9 sub basins are under very shallow and shallow soils exists in the upper parts of 13,14,15,16,17 and 18 sub basins. The escarpments and plateau spurs and high drainage density and steep to very steep slopes are the causative factors for shallow soils. The soils in the upper and lower sector of 5,8,13,16 and 19 sub basins are under moderately deep. The soils in lower parts of narrow valleys and main valley floor are predominantly under deep. The less drainage density, level slopes and longer flow of surface water are the influencing factors for existing deep soils. In the main valley floor soils are predominantly fluvial in nature and deep to very deep under well drained condition.

Soil Drainage

The analysis of soil drainage reveals that soils in dissected ridges, isolate mounds and linear ridges of the upper parts of 5,8,13,14,15,17 and 18th sub basins are very shallow to shallow, excessively drained and underlain by weathered basalt. The escarpments with steep to very steep slopes, have shallow and excessive drained soils. The soils in the parts of plateau spurs, dissected ridges, narrow valleys, middle and lower sectors and main valley floors are under well drained condition. The soils in the middle and lower sectors in the lower parts of 9,11 and

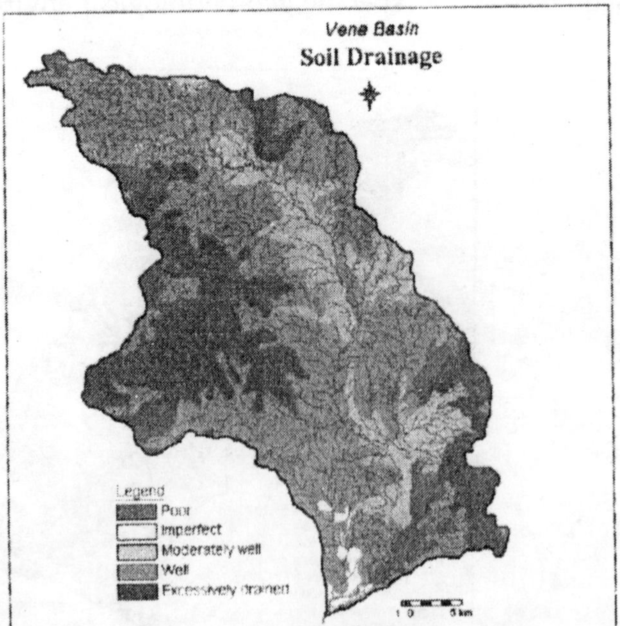

16 sub basins are under moderately well drained condition. The poor drained soils are mostly in unclassified area of the basin. The soils in flooding zone of 19th sub basin are under imperfect drainage because of its local topography and very less drainage density.

Available Water Holding Capacity

The Available Water holding Capacity (AWC) of the soils is dependent upon the intensity of rainfall, morphometry, infiltration, permeability, depth and volume of the soils. The AWC helps in deciding the length of growing period which is essential parameter for crop planning in an area more particularly under rained conditions. In the present study the analysis of soil AWC reveals that the zone of very low soil AWC (50-100) is in association with high drainage density, high stream frequency and steep slopes in the sub basins of 1 and 2. These prevailing conditions affect the rate of infiltration of 4,5,8,11,14,15,17 and 18 sub basins the soil AWC (100-150) is medium and it coincides with moderate drainage density, stream frequency, high bifurcation ratio and moderate elongation ratio. The high soil AWC (150-200) zone is mainly under unclassified area with low drainage density, moderate slopes and good vegetation cover. The very high soil AWC (>200) zone is in association with very low drainage density, low stream frequency and low texture ratio in the sub basins of 12 and 19. The lower part, the majority of sub basins have very high soil AWC conditions.

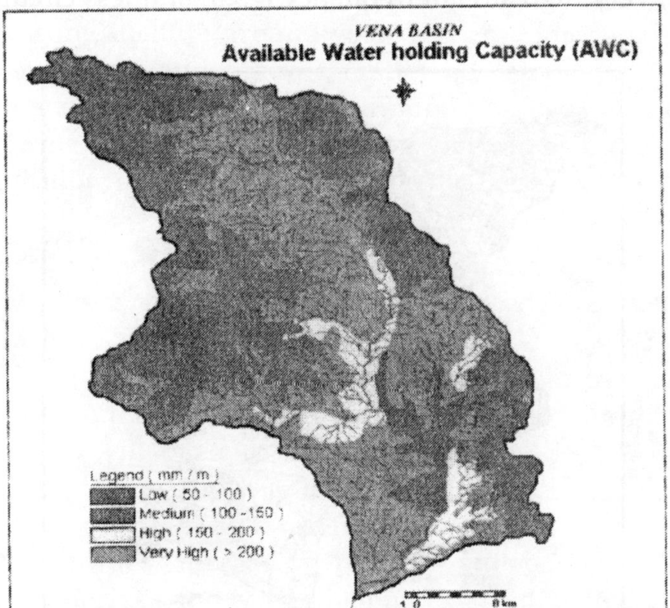

Eroded Lands

Based on the slope gradient, morphometry, broad soil physical characteristics, vegetation cover and image characteristics four categories of eroded lands i.e., very severe, severe, moderate and nil to slight were delineated from the standard FCC imagery of IRS-ID, LISS-III data at 1:50,000 scale in conjunction with the collateral and ground truth information. The nil to slightly eroded lands are in association with very gently sloping lands with moderate drainage density, low form factor, flat surfaces with much incipient loss of surface soil, double cropped area in the lower parts of 8, 13 and 15 sub basins. The moderately eroded lands are in the unclassified area, which consists of very gentle to gentle sloping lands with moderate vegetal cover and low drainage density. The severely eroded lands are in association with gently to moderately sloping lands, escarpments, plateau spurs and upper plateau plains with occasional gullies and rills, exposed to place showing sub-soil horizon and weathered parent material. These lands occur in the sub basins having high drainage density (>2.0), high bifurcation ratio (>4.0), high form factor (>0.4) and high texture ratio (>2.5). These factors are manifestation of the underlying lithology, moderate to high runoff and sparse vegetation cover. The very severely eroded lands are in the areas of dissected to undulating topography with steep to very steep slopes, at places very less vegetation cover associated with gullies, mainly covered with waste lands and occasional rock outcrops. The severe eroded lands are in association with very high drainage density (>2.5), very high stream frequency (>3.0) and less circulatory ratio (<0.5).

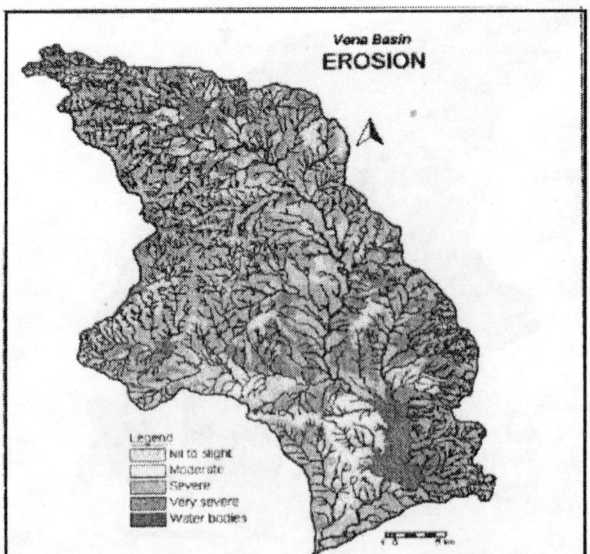

Groundwater Exploration and Recharge Site Selection

Lineament Density Map

In a hard rock area, the structurally controlled lineaments act as conduits for ground water movement (Anbazhagan and Ramasamy, 2006). Lineament analysis for ground water exploration in basaltic terrain has considerable importance, as joints and fractures serve as conduits for movement of ground water (Saraf and Choudhury, 1998). The lineament map (Fig. 1) is prepared by using topographic sheets, drainage maps and satellite imageries. Field check was also conducted for checking the conformity of the lineaments as geologic features of interest.. In the GIS analysis, we cannot use such lineament maps directly, which should be converted into lineament density map. To do so, from such lineament data, the study area was divided into square grids of 1 sq. km and the total lengths of

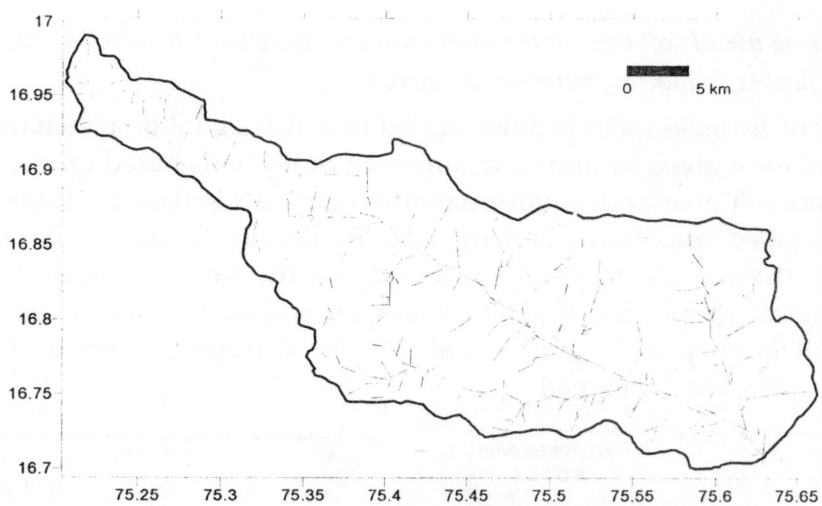

Fig. 1. Lineament map over laid by grid map for developing lineament density map

lineament falling in each grid were measured in order to determine the lineament density in km/sq. km, further these were plotted on the respective grid centre and contoured. Lineaments in combination with the drainage network may be used to locate the different recharge structures (Fig. 2).

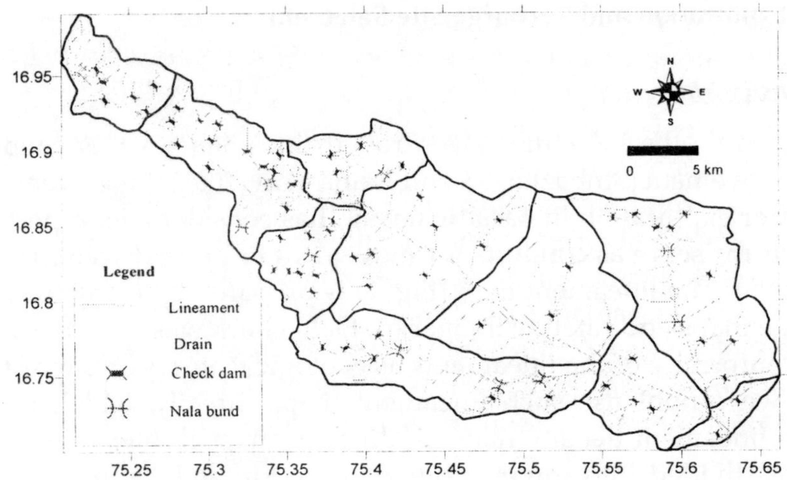

Fig. 2. Map showing the proposed sites for various water harvesting and recharge structures in the study area.

Method 2 : Making use of soft copy of the Karnataka soils developed by NBSS & LUP, Bangalore and further conducting detailed soil survey

The soft copy of *Karnataka soils* is down scaled to a district, taluk, panchayat and cadastral level for a given locality .Classification of the soils based on the above information doesn't give critical information on detailed soil units. Therefore it is essential detailed soil survey and the information on different soils units is brought out thematically. Further, based on the soil types and other geomorphological properties, thematic maps on soil and water conservation practices, soil quality units, soil suitability for different crops and other management units are developed.

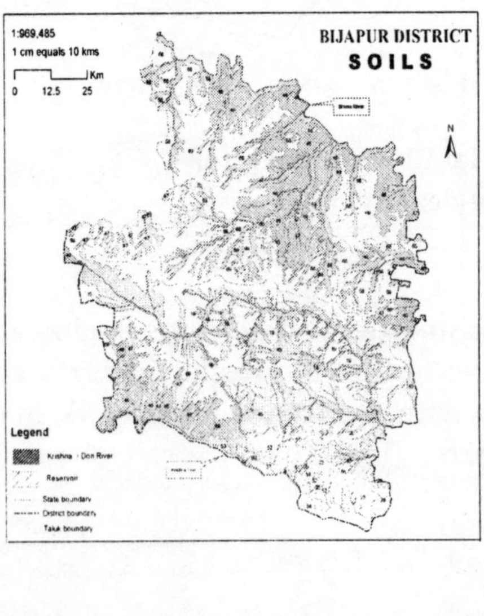

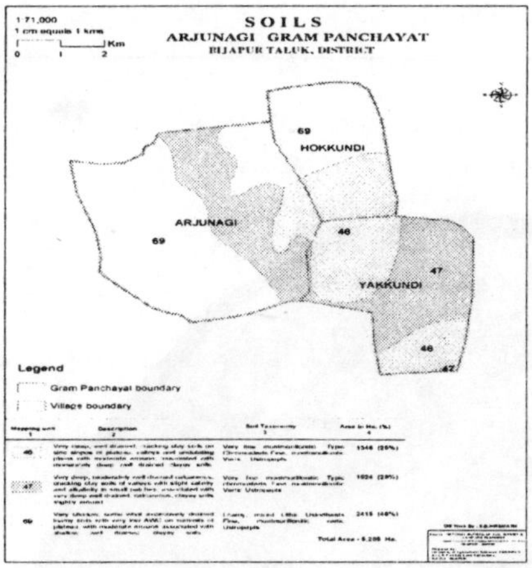
SOILS
ARJUNAGI GRAM PANCHAYAT
BIJAPUR TALUK, DISTRICT
HOKKUNDI
ARJUNAGI
YAKKUNDI
Legend
Gram Panchayat boundary
Village boundary

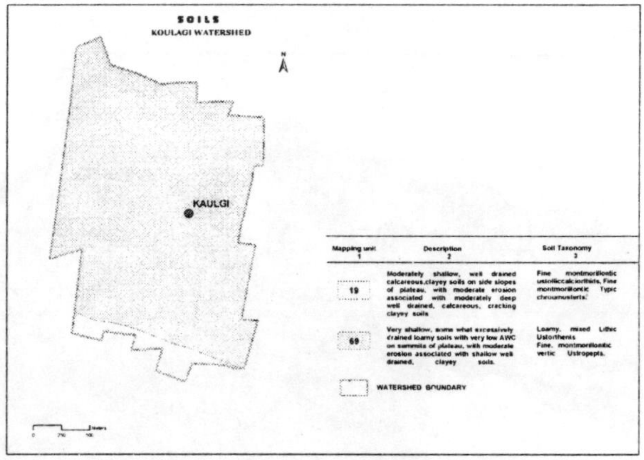
SOILS
KOULAGI WATERSHED
KAULGI
Mapping unit
Description
Soil Taxonomy
WATERSHED BOUNDARY

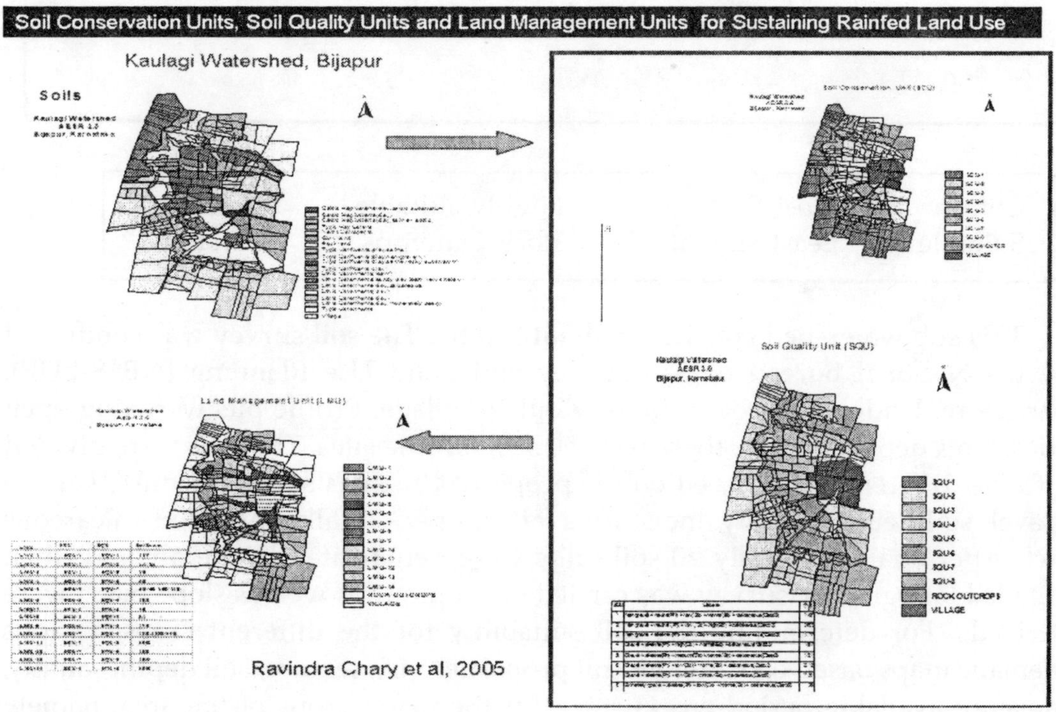
Soil Conservation Units, Soil Quality Units and Land Management Units for Sustaining Rainfed Land Use
Kaulagi Watershed, Bijapur
Soils
Soil Conservation Unit (SCU)
Soil Quality Unit (SQU)
Land Management Unit (LMU)
SQU-1
SQU-2
SQU-3
SQU-4
SQU-5
SQU-6
SQU-7
SQU-8
ROCK OUTCROPS
VILLAGE
Ravindra Chary et al, 2005

Soil site suitability for the different crops (Case study)

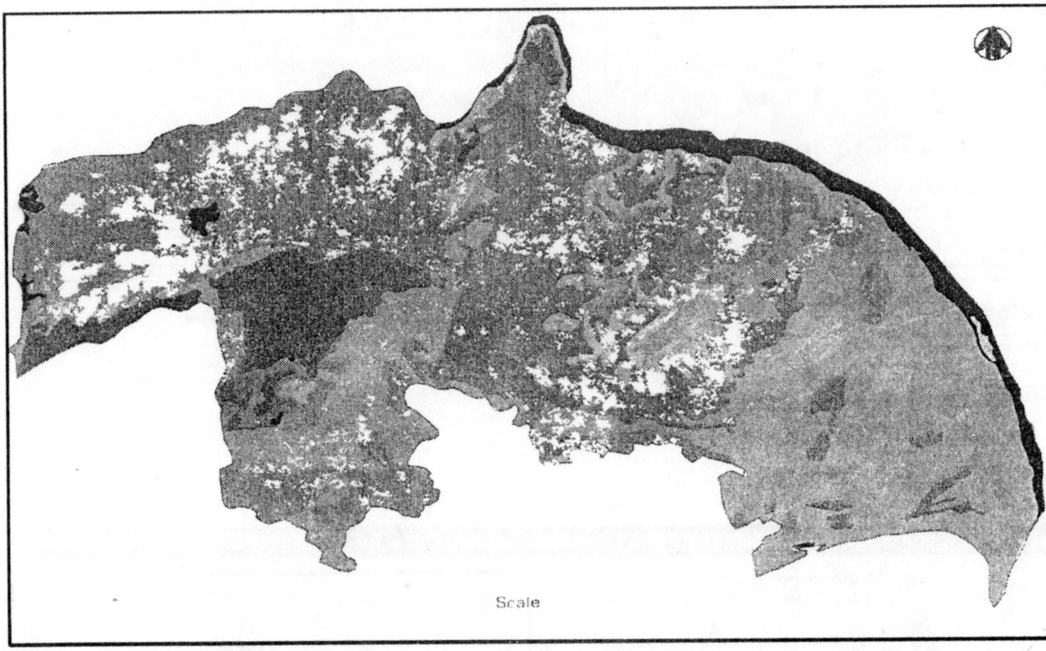

Cotton	Most Suitable	Highly suitable	Moderately
Suitable	Least suitable	Rock Outcrops	Forest

This sub watershed consists of about 350 ha. The soil survey was conducted by the National bureau of Soil Survey and Land Use Planning (NBSS-LUP), Bangalore. Under the project, in the Koulagi village, Profile pits were dug open atleast to a depth of 1.5 m, there were 54 such profile sites. The soils were divided into the different types based on the properties such as surface texture, surface gravel, soil depth, salinity, moisture available period, alkalinity and calcareous etc. After analysis, totally 20 soil units were demarcated in the area. Further, detailed topographic survey was carried out contour as well as slope maps were derived. For determining the soil suitability for the different crops, various thematic maps based on different soil properties viz., texture, soil depth, salinity, moisture available period etc. Further, for the major crops of the area, namely sunflower the soil suitability criteria developed by NBSS- LUP, Bangalore was used. For the analysis, four suitability classes were used, class S1- highly suitable, class S2- moderately suitable. Class S-3 marginally suitable and class n non suitable.

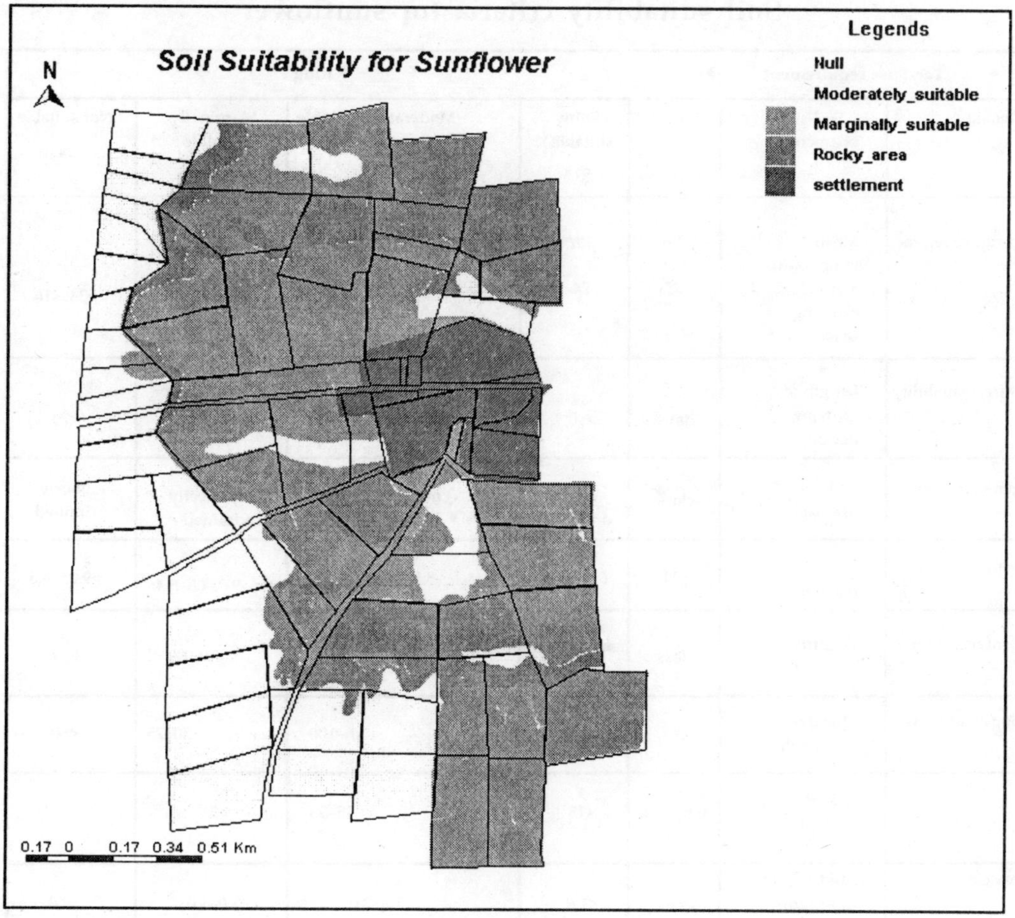

Characterization of Water Resource

Sky rainwater, surface rainwater and ground water resources together account for total water resources of a locality. Water is the key issue accelerating the helical wheel of crop production process , but its speed is frequently being distorted because of irregular rains both in quantity and quality spatially and temporarily. This is very much evident when we observe food production scenario of the country across years.

It's of para-mount important in characterization of rainwater for a given locality to achieve this it's required to :

- Identify the areas prone for rainfall changes.
- Asses the impacts of rainfall variability at District/Tehsil/ rain gauge stations and area affected (identification of hot spots)

Soil suitability criteria for sunflower

← Land use requirement →			← Rating →			
Land quality	Soil-site characteristic	Unit	Highly suitable S1	Moderately suitable S2	Marginally suitable S3	Not suitable N
Temperature regime	Mean temperature in growing seasn	°C	24–30	30–34; 20–24	34–38; 16–20	>38; <16
Moisture availability	Length of growing period	days	>90	80–90	70–80	<70
Oxygen availability to roots	Soil drainage	class	Well drained	moderately well drained	imperfectly drained	poorly drained
Nutrient availability	Soil reaction	pH	6.5–8.0	8.1–8.5; 5.5–6.4	8.6–9.0; 4.5–5.4	>9.0; <4.5
Nutrient retention	Texture	class	l, cl, sil, sc	scl, sic, c	c >60%, sl	ls, s
Rooting conditions	Effective soil depth	cm	>100	76–100	50–75	<50
	Gravel content	% by vol.	<15	15–35	>35	
Soil toxicity	Salinity (EC saturation extract)	dS m^{-1}	<1.0	1.0–2.0	2.0–4.0	>4.0
	Sodicity (ESP)	%	<10	10–15	>15	
Erosion hazard	Slope	%	<3	3–5	5–10	>10

- Asses the impact of rainfall and thermal regime variations on soil carbon storage and landuse.
- Studies on pattern of drought intensity and development of area specific management strategies.
- A re-look at crop improvement and water management strategies variability

A Case Study

Keeping in mind the above aspects the historical rainfall of Bijapur(100 years) was analyzed critically and the information as variations in annual, seasonal both

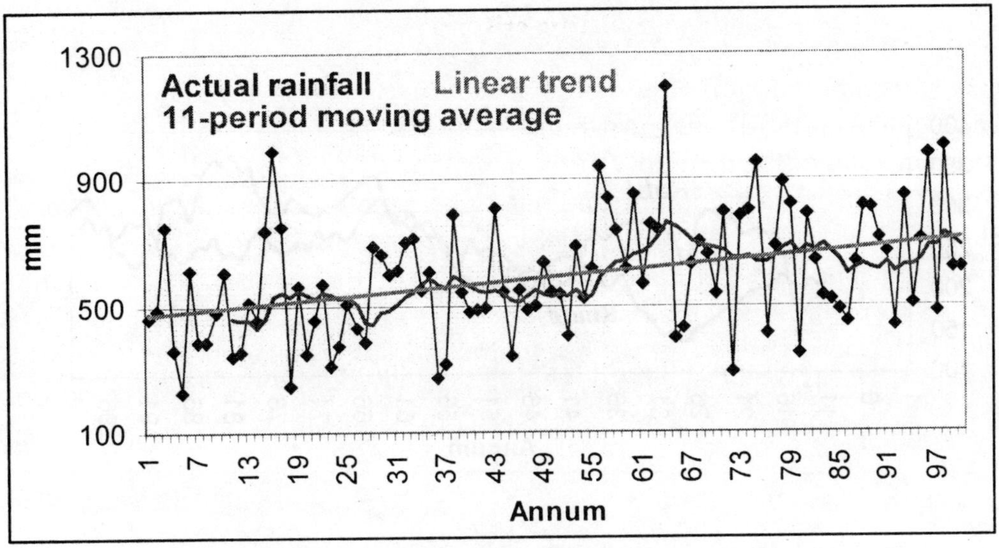

Fig. 4.3. Annual rainfall at Bijapur during 1901-2000

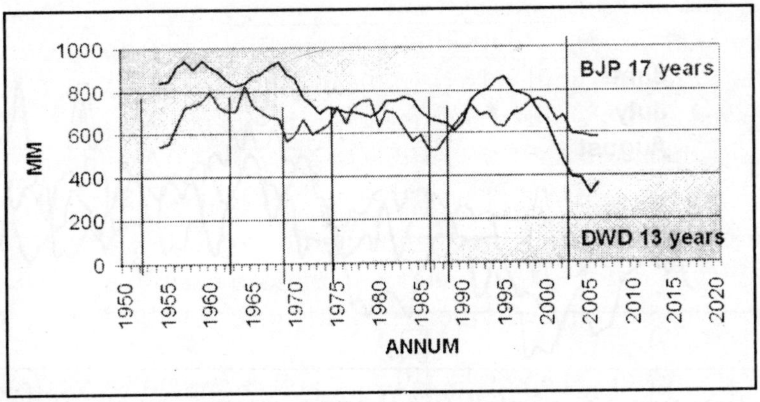

Fig. 4.4. Comparison of annual rainfall at Bijapur and Dharwad: Return of low rainfall spochs

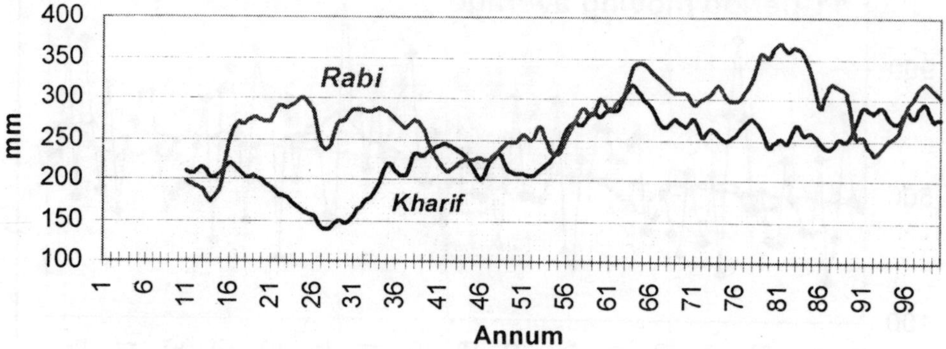

Fig. 4.5. Secular changes in seasonal rainfall at Bijapur

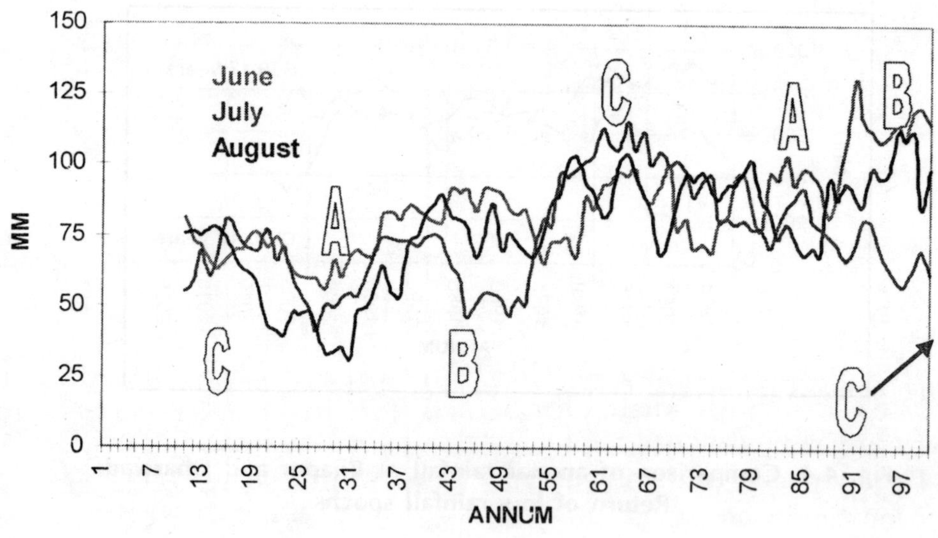

Fig. 4.6. Rainfall during Kharif months at Bijapur

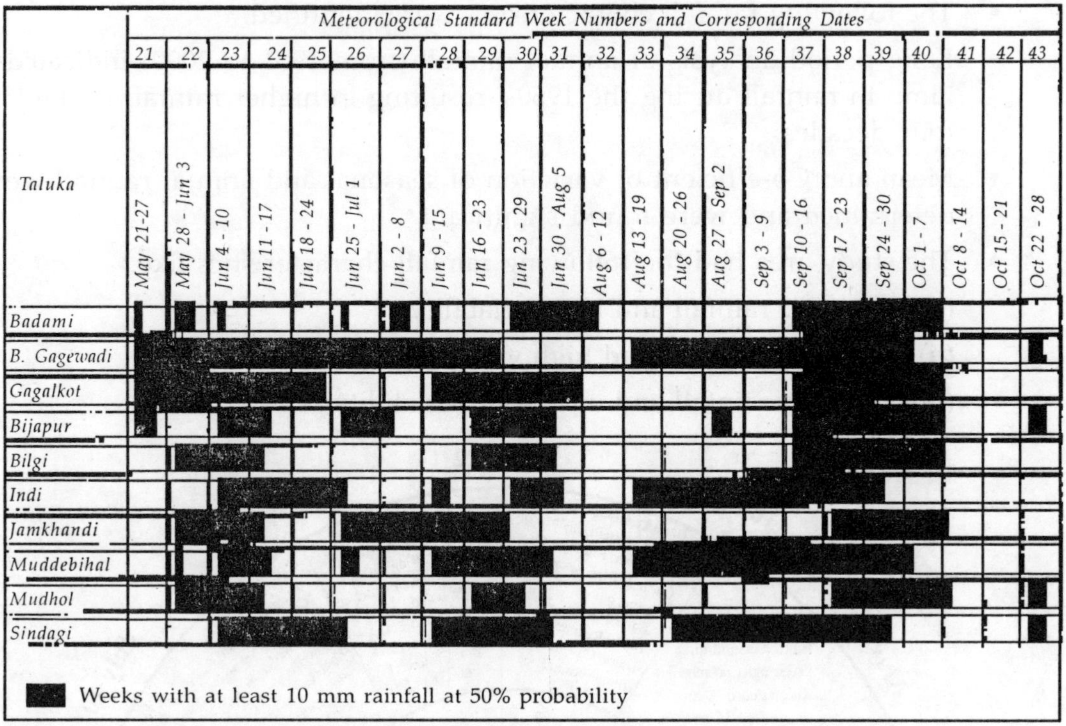

Schematic diagram showing weeks with at least 10.0 mm in weeks/groups of weeks at different talukas of Bijapur & Bagalkot districts

inter and intra seasonal and return of low/high rainfall epochs have been established. The information is presented below.

Rainwater

- The probability of occurrence of deficit rainfall is 76 per cent and its spatial and temporal variability ranges from 31 to 64 per cent.
- High insolation (18.9-21.0 MJ/M2/day) and desiccating wind (20 km/hr) cause annual potential evaporation of 1574 mm that far exceeds rainfall by 988 mm. The aridity index is of 62 per cent.
- Drought generally occurs twice in every five years, while the severe droughts that persist for 3-4 years occur nearly once in 30 years.
- Since 16th century, drought has occurred 423 times. Out of them severe drought has occurred for 73 times.

Rainfall

- The mean annual rainfall over the study area varied from 518 mm in the extreme Southern Taluka of Badami to 639 mm in the Central region

- The following rainfall characteristics were identified.
- Long period analysis of annual rainfall (from 1901 to 2000) indicated a jump in rainfall during the 1960s, resulting in higher rainfall in the last four decades.
- Mean and Co-efficient of variation of seasonal and annual rainfall were determined and were found significant.
- The study area had the following rainfall characteristics like:

  (i) Increased rainfall and low variability.

  (ii) Decreased rainfall and high variability and

  (iii) Increased rainfall and increased variability

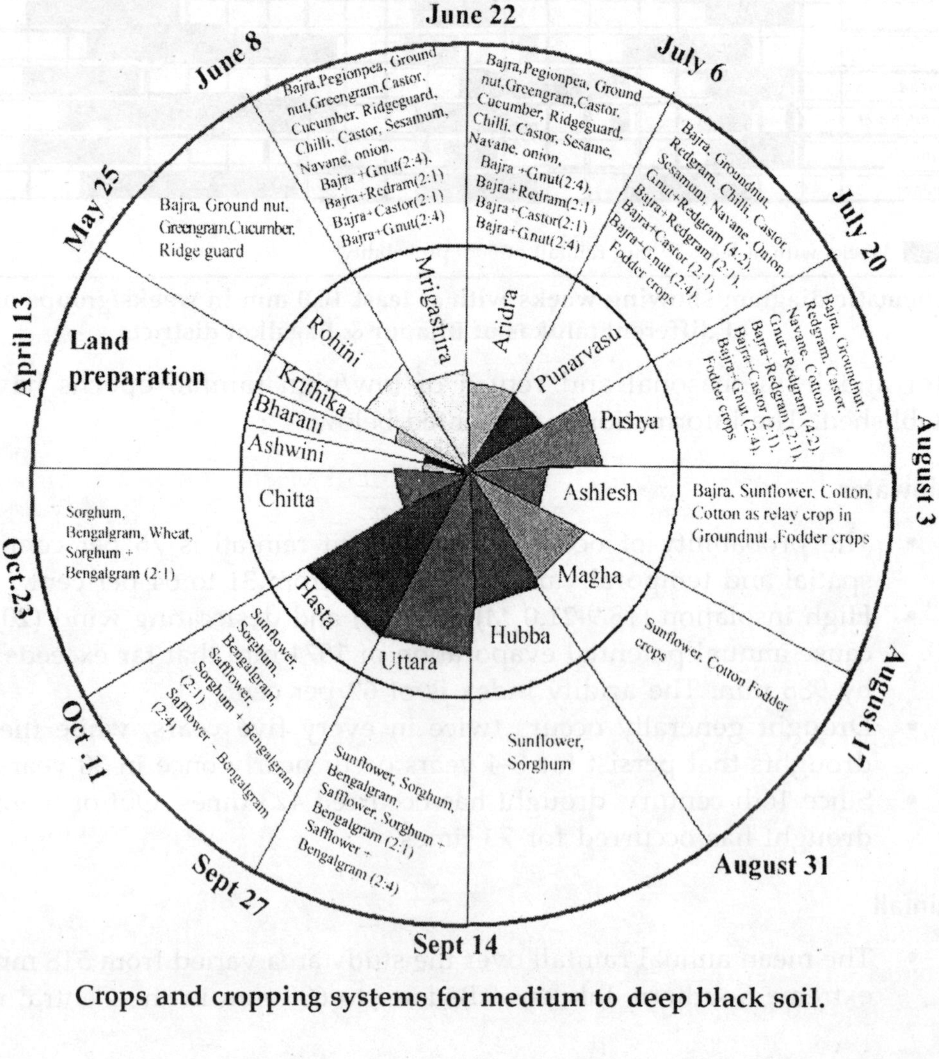

Crops and cropping systems for medium to deep black soil.

- In addition to the climatological variations, the spatial variations in intra-seasonal fluctuations for all the talukas of two district Bijapur & Bagalkot have been identified.
- This information is important for effective micro-level planning and implementation

Management Practices

The potential soil and water conservation structures recommended for cultivable lands are

- Contour bunds with land shaping.
- Graded bunds with land shaping and strip leveling.

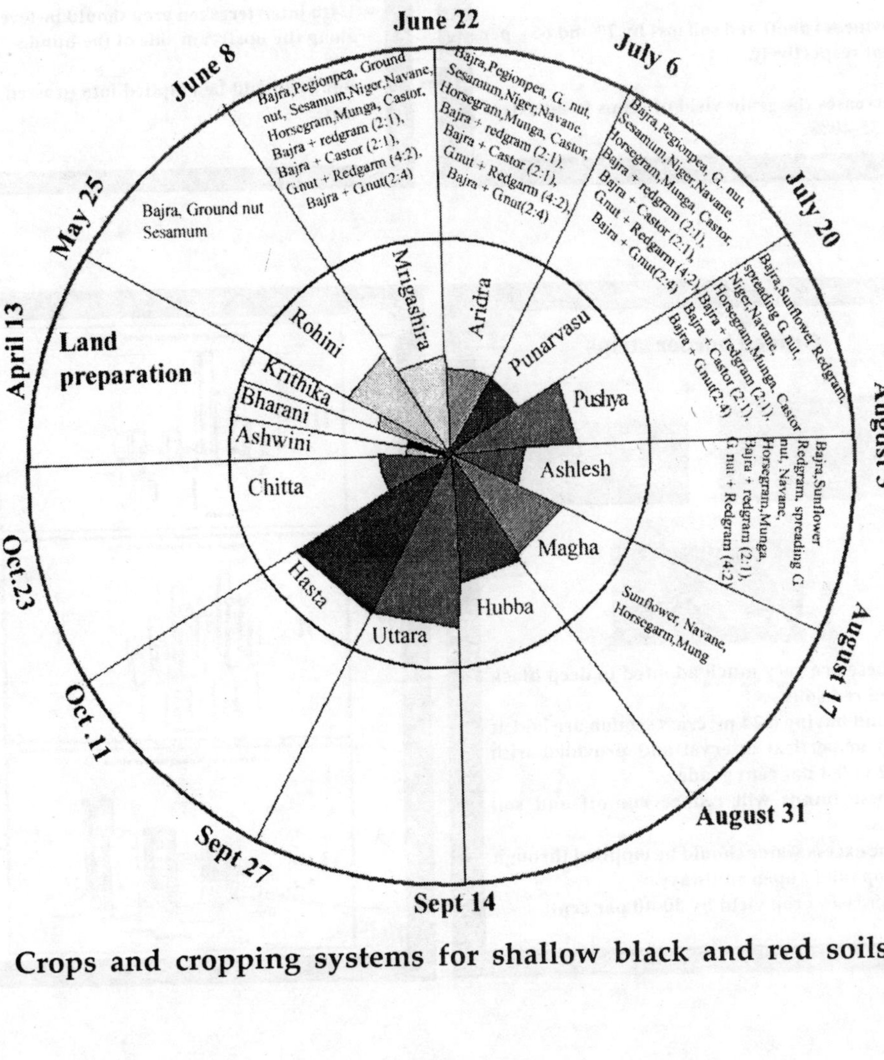

Crops and cropping systems for shallow black and red soils.

Contour bunds

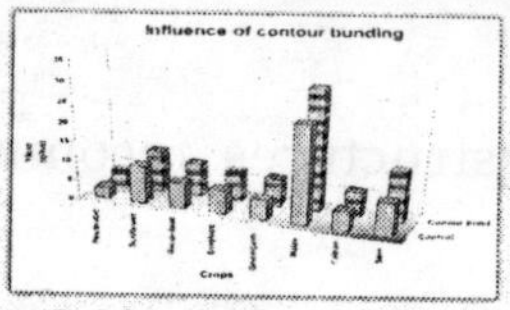

- Recommended for shallow black and red soils.
- Bunds are constructed on contour key lines at 1 to 1.5 m vertical interval with 0.81 to 1.4 m^2 cross section.
- Reduces runoff and soil loss by 70 and 65 per cent respectively.
- Increases the grain yield of crops to an extent of 35-40%.

Graded bund with strip leveling

- Recommended for deep black soils and shallow to medium red soils.
- Construct bunds of 0.85 m^2 cross section at 1 to 1.5 m vertical interval on 0.2 to 0.4% grade.
- 1/4th inter-terraced area should be leveled all along the upstream side of the bund.
- Bunds should be emptied into grassed water ways.

Graded border strips

- These are very much adopted in deep black and red soils.
- Bund having 0.24 m^2 cross section are laid at 0.3 m vertical interval and provided with 0.2 to 0.4 per cent grade.
- These bunds will reduce run-off and soil loss.
- The excess water should be emptied through drop inlet / open spillways.
- Increases crop yield by 30-40 per cent.

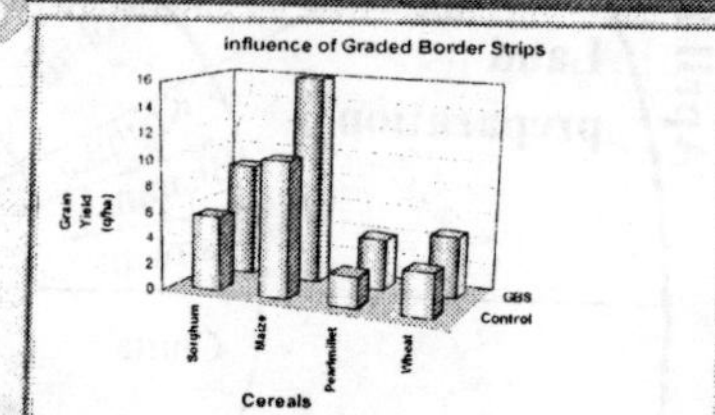

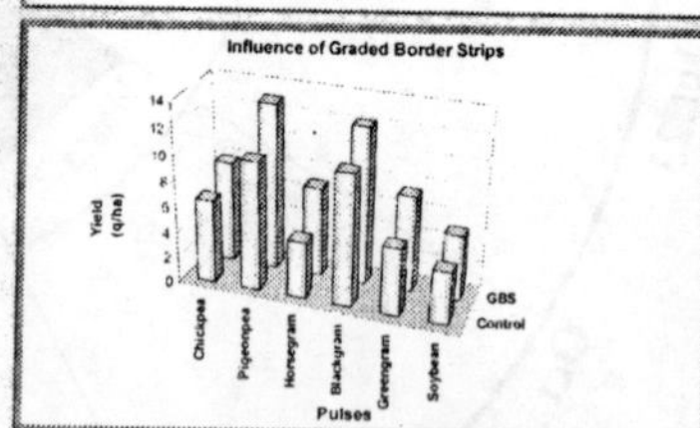

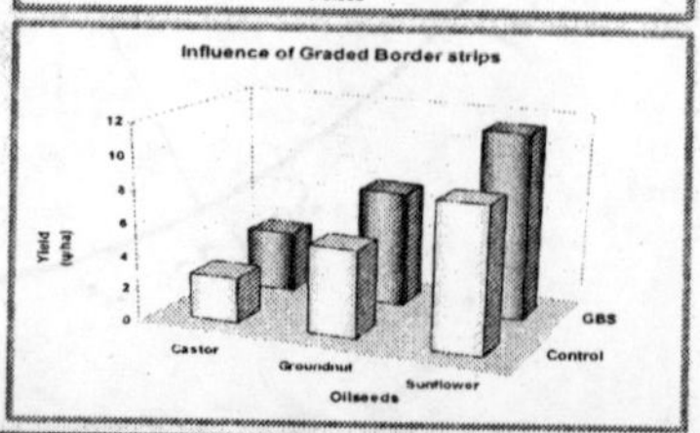

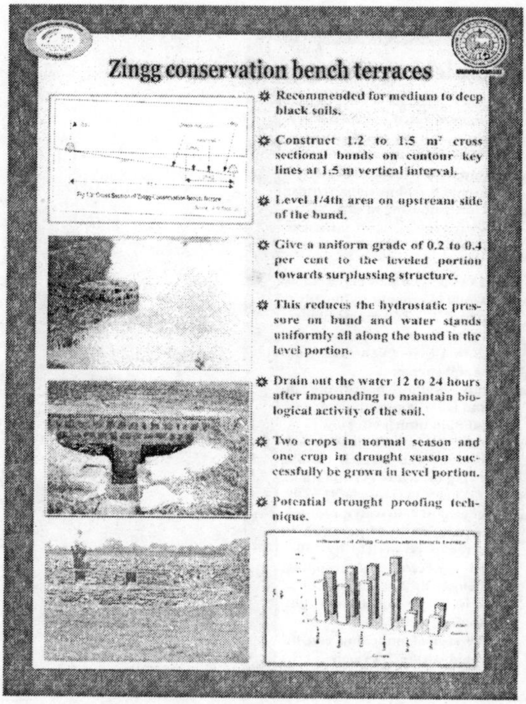

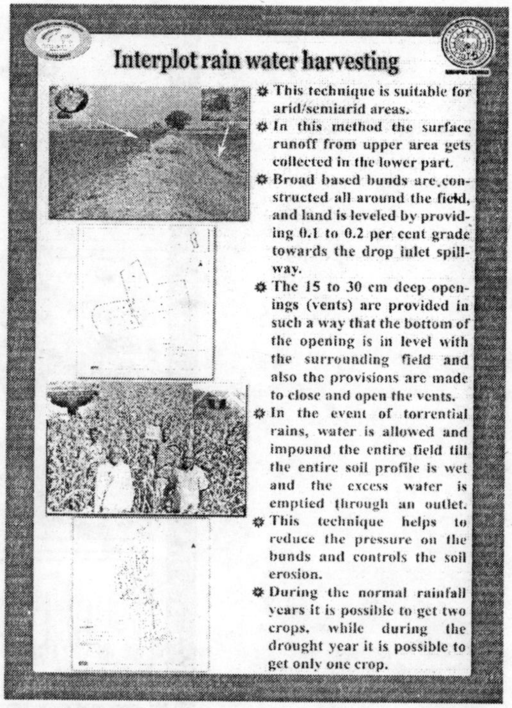

Interplot rainwater harvesting technique has shown to conserve the rain water and soil by 95 to 98 % and increase the grain yield by 200 to 300 %. It is one of the most drought proofing mechanisms.

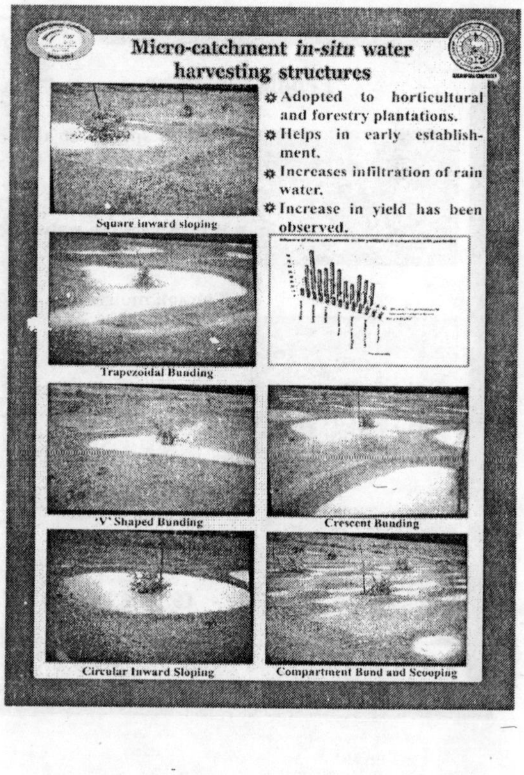

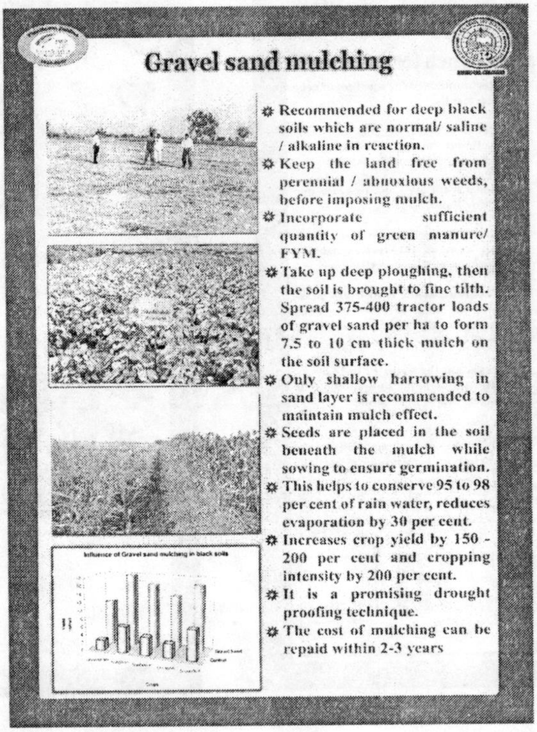
Gravel sand mulching
Recommended for deep black soils which are normal/ saline / alkaline in reaction.
Keep the land free from perennial / abuoxious weeds, before imposing mulch.
Incorporate sufficient quantity of green manure/ FYM.
Take up deep ploughing, then the soil is brought to fine tilth. Spread 375-400 tractor loads of gravel sand per ha to form 7.5 to 10 cm thick mulch on the soil surface.
Only shallow harrowing in sand layer is recommended to maintain mulch effect.
Seeds are placed in the soil beneath the mulch while sowing to ensure germination.
This helps to conserve 95 to 98 per cent of rain water, reduces evaporation by 30 per cent.
Increases crop yield by 150 - 200 per cent and cropping intensity by 200 per cent.
It is a promising drought proofing technique.
The cost of mulching can be repaid within 2-3 years
Influence of Gravel sand mulching in black soils

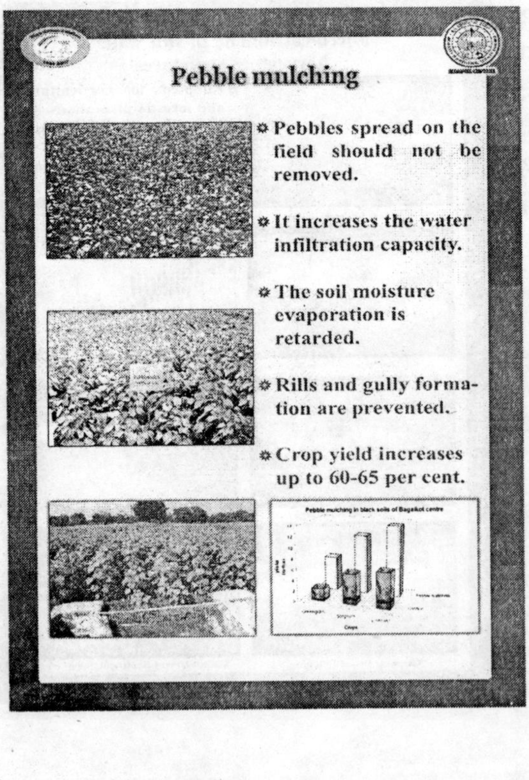
Pebble mulching
Pebbles spread on the field should not be removed.
It increases the water infiltration capacity.
The soil moisture evaporation is retarded.
Rills and gully formation are prevented.
Crop yield increases up to 60-65 per cent.
Pebble mulching in black soils of Bagalkot centre

Set furrow cultivation
(Tank silt applied)
Green leaf manure crops along the field boundary
Set furrow
Cutting of green leaf manure
Application of green leaf manure + tank silt
Filled in furrows with green leaf manure + tank silt
Sunflower + horsegram in set furrows

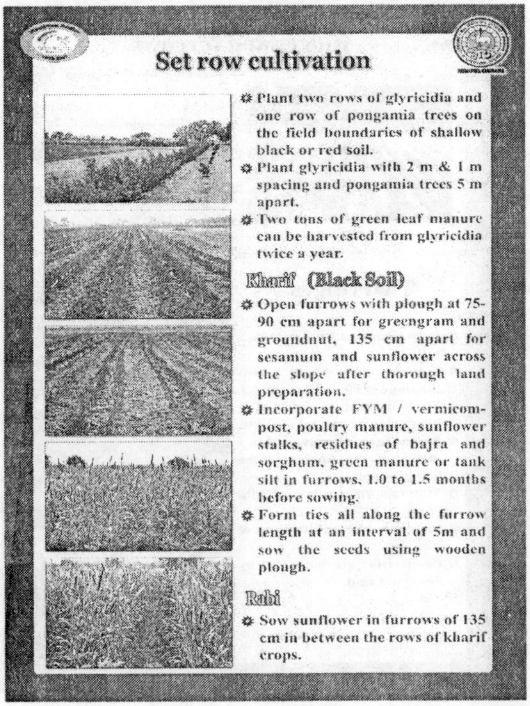
Set row cultivation
Plant two rows of glyricidia and one row of pongamia trees on the field boundaries of shallow black or red soil.
Plant glyricidia with 2 m & 1 m spacing and pongamia trees 5 m apart.
Two tons of green leaf manure can be harvested from glyricidia twice a year.
Kharif (Black Soil)
Open furrows with plough at 75-90 cm apart for greengram and groundnut, 135 cm apart for sesamum and sunflower across the slope after thorough land preparation.
Incorporate FYM / vermicompost, poultry manure, sunflower stalks, residues of bajra and sorghum, green manure or tank silt in furrows, 1.0 to 1.5 months before sowing.
Form ties all along the furrow length at an interval of 5m and sow the seeds using wooden plough.
Rabi
Sow sunflower in furrows of 135 cm in between the rows of kharif crops.

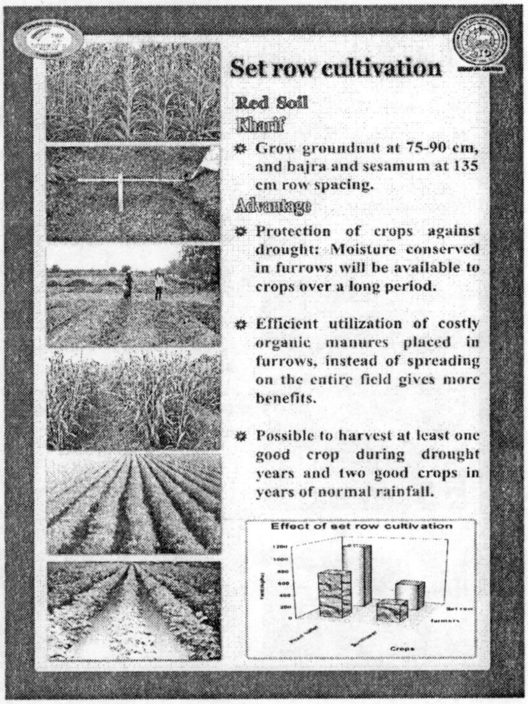
Set row cultivation
Red Soil
Kharif
Grow groundnut at 75-90 cm, and bajra and sesamum at 135 cm row spacing.
Advantage
Protection of crops against drought: Moisture conserved in furrows will be available to crops over a long period.
Efficient utilization of costly organic manures placed in furrows, instead of spreading on the entire field gives more benefits.
Possible to harvest at least one good crop during drought years and two good crops in years of normal rainfall.
Effect of set row cultivation
Crops

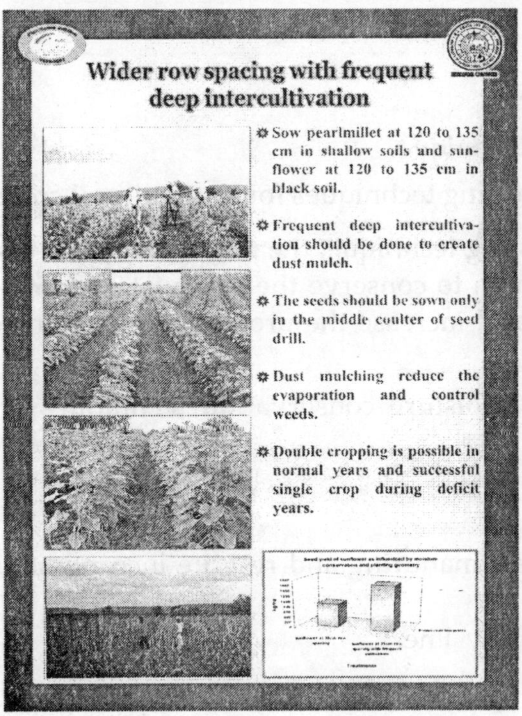
Wider row spacing with frequent deep intercultivation
Sow pearlmillet at 120 to 135 cm in shallow soils and sunflower at 120 to 135 cm in black soil.
Frequent deep intercultivation should be done to create dust mulch.
The seeds should be sown only in the middle coulter of seed drill.
Dust mulching reduce the evaporation and control weeds.
Double cropping is possible in normal years and successful single crop during deficit years.

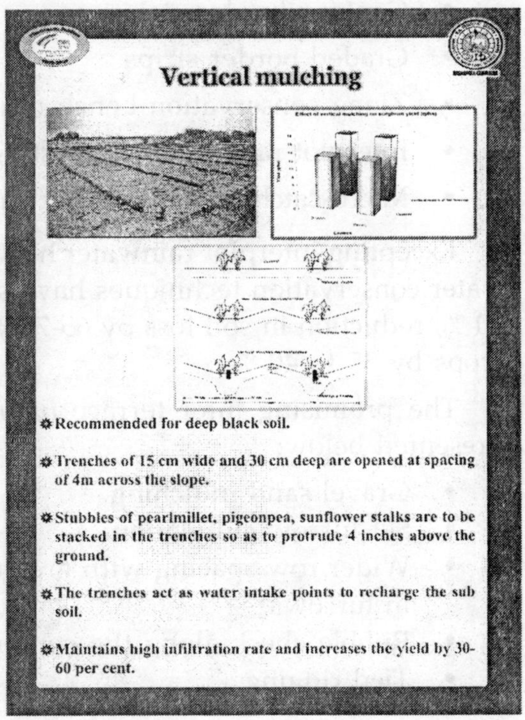
Vertical mulching
Recommended for deep black soil.
Trenches of 15 cm wide and 30 cm deep are opened at spacing of 4m across the slope.
Stubbles of pearlmillet, pigeonpea, sunflower stalks are to be stacked in the trenches so as to protrude 4 inches above the ground.
The trenches act as water intake points to recharge the sub soil.
Maintains high infiltration rate and increases the yield by 30-60 per cent.

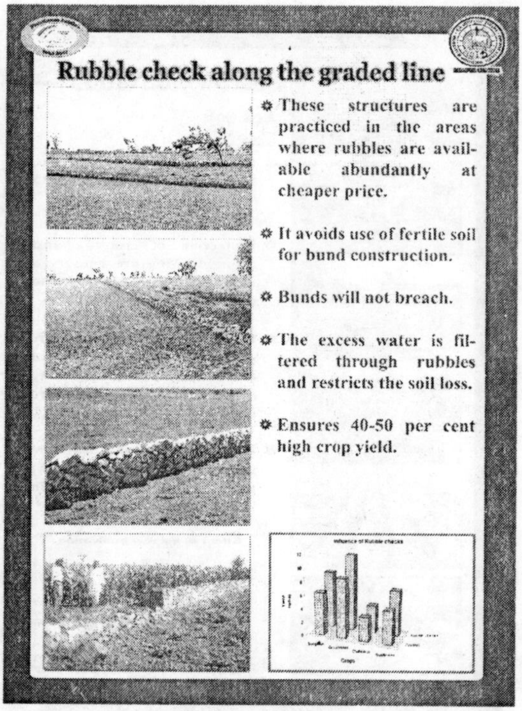

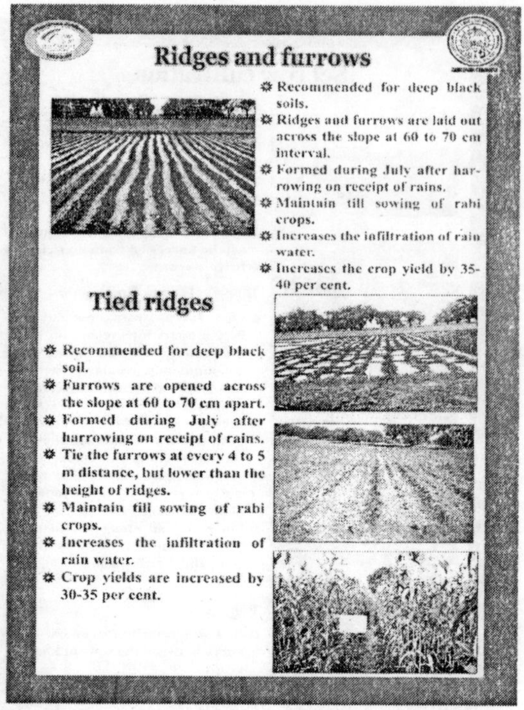

- Contour border strips.
- Graded border strips.
- Zingg conservation bench terraces.
- Interplot rain water harvesting technique.
- Micro-catchment rain water harvesting techniques for horticultural crops.

Excepting interplot rainwater harvesting technique, the above listed soil and water conservation techniques have shown to conserve the rain water by 60 to 70 %, reduction in soil loss by 65-75 % and increase the productivity of rainfed crops by 35 to 40 %.

The promising inter terrace *in situ* moisture conservation techniques are presented below:

- Gravel sand mulching.
- Set furrow cultivation.
- Wider row spacing with green leaf manuring and residue incorporation in furrows.
- Rubble check along the contour key line.
- Tied ridging.

Among the five technologies listed above, gravel sand mulch has shown to harvest the rain water effectively (95 to 98 %) and retains it for a considerable period of time. Has shown to increase the crop yield by 200 to 300 % and cropping intensity by 150 to 200 % even during drought years. It requires higher initial investment.

Set furrow cultivation and wider row spacing with conservation furrow practices not only provide sustainable higher income even during drought years but also improve the physical, chemical and biological proportions of shallow soils which occupy considerable area in dryland agro-eco-system.

Farm Pond Water Utilization

Research conducted in several areas of drylands in the country and also else where have revealed the feasibility of harvesting and storing the excess run-off in dugout farm ponds and reutilizing the same for providing supplemental irrigations to crops during prolonged periods of drought is one of the techniques of enhancing the productivity of dryland crops.

For enhancing the water use efficiently of farm pond water, adoption of sprinkler or drip systems are more promising methods of irrigation.

The pertinent question more frequently put forth by the farming community is that how these two systems of irrigation will be relevant with respect to their utility round the year and durability of the system during off season.

The initial investment on any one of these systems will be very high. (Rs 20 to 25 thousand/acre for drip & 15 to 18 thousand per acre for sprinkler)

In order to keep the above mentioned pressurized methods of irrigation functioning even during off season, it is inevitable on the port of the user to link the system to a permanent water source which may be of open well or bore well.

Further these sources may be suitably located down below the farm pond so that they get recharged.

Selection of Crops

The harvested run-off in farm ponds is costly and scarce commodity in dryland agriculture. The stored water should be used most judiciously.

It is not always necessary that the stored water should be used to provide supplemental irrigations to the standing crops during stress period.

For enhancing the productivity of harvested rain water and also to realize sizeable income and year round employment to completely meet out the livelihood requirement of the Marginal or Small farmer it is of great important to select a promising integrated cropping system involving fruit, flower, olericulture crops and other important food crops having cereals, pulses and oilseed, will be a good combination.

Conclusion

The present challenge of dryland agriculture is to manage spatial and temporal variability associated with rainwater and soil resources by applying suitable technology and principles and to relocate both the resources appropriately with other inputs and practices to localized condition within a field to achieve improved crop production and environment quality.

5

Application of Remote Sensing and GIS Techniques for Groundwater Recharge Site Selection in Hard Rock Areas—A Case Study from South India

M.S. Shirahatti, M.V. Ranghswami, R. Sivasamy, S. Santana Bosu, M.V. Manjunath and M.B. Guled

The present study was taken up in the upper Don river basin, Bijapur district of Karnataka state. The physiography is gently sloping, the normal annual rainfall of Bijapur is 590 mm, while the monthly reference evapotranspiration (ETo) is in the range of 82.77 to 258.23 mm. The major part of the upper Don watershed is occupied by basalt and other trap intrusions, while, black cotton soil is dominant type of soil.In this study, nine different thematic layers viz., geology, soil, land use / cover, water table fluctuation, depth of bed rock, slope, drainage density, lineament density and geomorphology were used as the input for identifying the recharge sites. For the development of land use/cover map, the IRS ID LISS-III digital data was used, while Shuttle Radar Topography Mission (SRTM) data was used for developing the slope map. To meet the specific objectives, the multi criteria evaluation (MCE) was carried out. The model was evaluated with the bore well yield data and it was found that, suitable zones demarcated by the model was matched better with the well yield data. The output of the WLC model revealed that 0.93, 48.5, 50 and 0.57 percentage of the total areas were occupied by the poor, moderate, good and excellent recharge suitability zones respectively

Introduction

Groundwater is a dynamic and replenish able natural resource but in hard rock terrains availability of groundwater is of limited extent. Occurrence of groundwater in such rocks is essentially confined to fractured and weathered horizons. In India, 65 per cent of the total geographical area is covered by hard rock formations with low porosity (less than 5 per cent) and very low (10^{-1} to 10^{-5} m/day) permeability (Saraf and Choudary, 1998 and Sahu, 2001). Therefore, efficient management and planning of groundwater in these areas is of the utmost importance. Proper scientific planning and management require immense data to make predictions of water availability.

Materials and Methods

Study Area

The present study was taken up in the upper Don river basin, Bijapur district of Karnataka state (Fig. 1) Its catchment is bounded by $10^{0}42'8''$ to $16^{0}59'26''$ N and $75^{0}12'24''$ to $75^{0}39'40''$ E. The major portion of the catchment falls in Bijapur taluka only. The physiography is gently sloping and there is the total relief of 143 m, normal rainfall of Bijapur is 590 mm and minimum and maximum temperatures recorded were 9.48^{0}C and 41.75^{0}C respectively. While the monthly reference evapotranspiration (ETo) was in the range of 82.77 to 258.23 mm. The major part of the upper Don watershed is occupied by basalt and other trap intrusions. Black cotton soil is dominant type of soil and it is followed by mixed loamy soil.

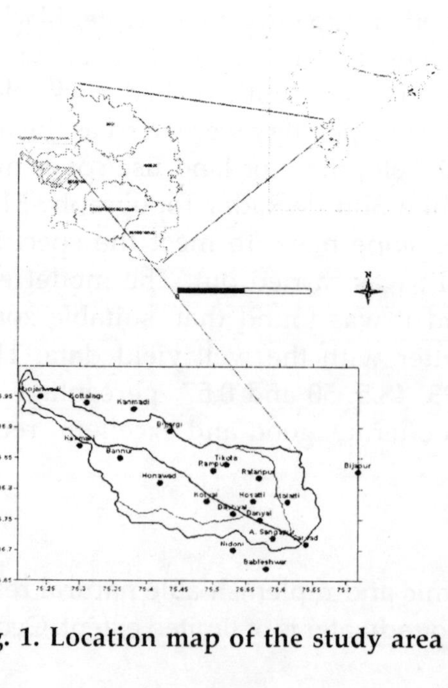

Fig. 1. Location map of the study area

Thematic Map Generation

In this study, nine different thematic layers were used. Among the nine layers, six layers viz., geology, soil, land use/cover, water table fluctuation, depth to bed rock and slope thematic maps were generated from the different data sources. For developing slope map, SRTM data was used. Other three thematic maps viz., drainage density, lineament density and geo morphology were derived from the above maps used. For the development of land use/cover map the IRS ID LISS-III digital data was used and by digital image processing, technique, totally eight different land use/land cover classes were demarcated in the study area. The high transformed divergence (TD) values among the classes were observed. It shows that these classes were spectrally separable. After developing different thematic maps, under the GIS environment geo data base was developed.

Modeling for Site Selection

To meet a specific objective, it is frequently the case that several criteria will need to be evaluated. Such a procedure is called Multi-Criteria Evaluation (Carver, 1991). Another term that is sometimes encountered for this is modeling. The artificial recharge process is presented in the form of flow diagram in Fig.2. density and geo morphology were derived from the above maps used. For the development of land use/cover map the IRS ID LISS-III digital data was used and by digital image processing, technique, totally eight different land use/land cover classes were demarcated in the study area. The high transformed divergence (TD) values among the classes were observed. It shows that these classes were spectrally separable. After developing different thematic maps, under the GIS environment geo data base was developed.

To meet a specific objective, it is frequently the case that several criteria will need to be evaluated. Such a procedure is called Multi-Criteria Evaluation (Carver, 1991). Another term that is sometimes encountered for this is modeling.

Weighted Linear Combination (WLC) Model or (Indexing Overlay)

Comparing weighted linear combination (WLC) models with Boolean Models executive routines, it is identified that WLC has more flexibility and ability for priority indication on spatial units of factor maps (Bonham Carter and G.F., 1991). In this type of models, it is possible to consider knowledge driven weighting. With respect to mentioned characteristics, this model is relatively simple and more realistic, hence this model was used for artificial recharge site selection process. With weighted linear combination (WLC), factors are combined by applying a weight to each followed by a summation of the results to yield of

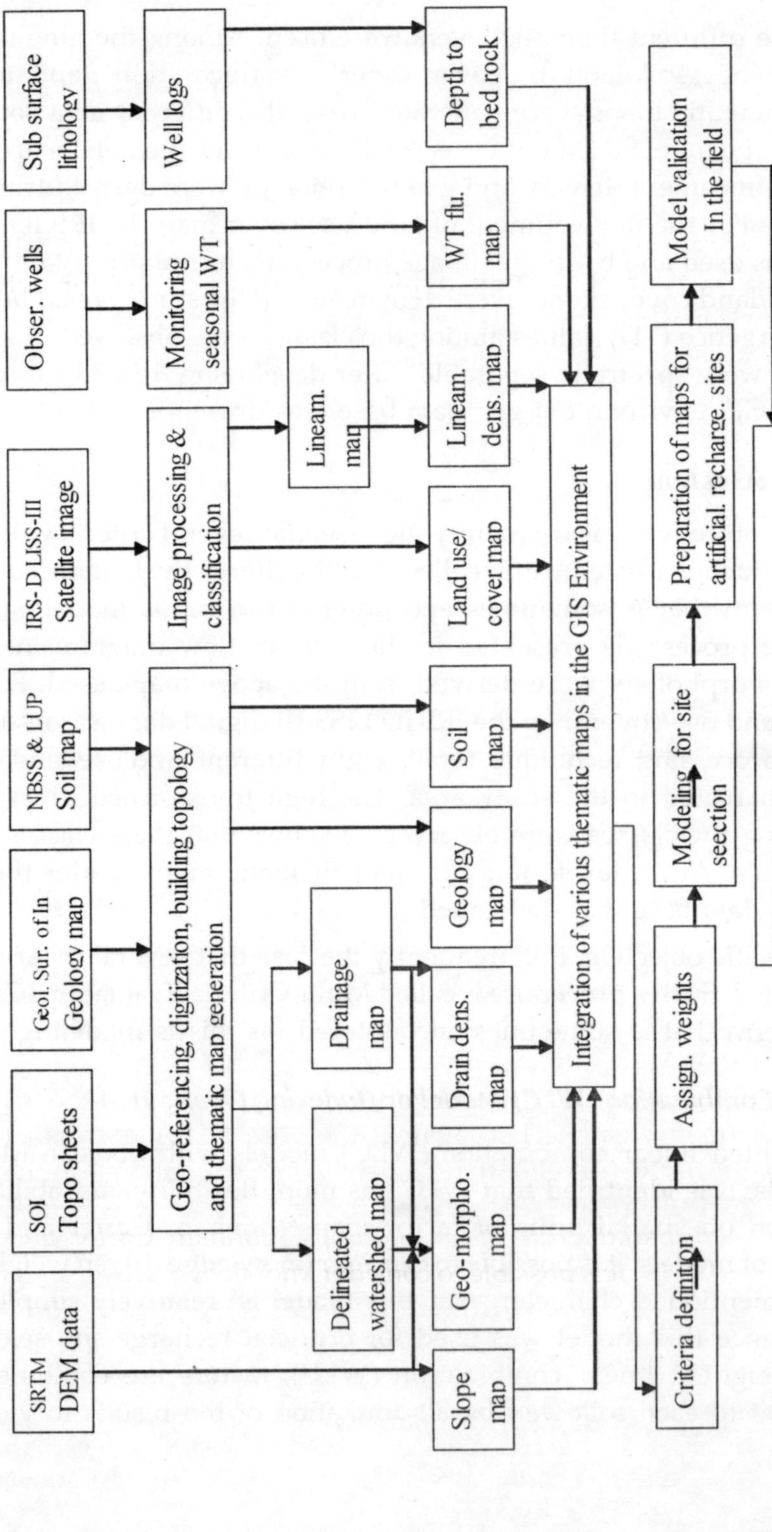

Fig. 2. Flow diagram showing the various steps involved in the artificial recharge site selection

suitability map (Eastman,2001) i.e.;

$$S= \Sigma\ wi\ xi \quad \ldots (1)$$

where ,

wi = The weight of ith factor map

xi = Criteria score of class of factor i

S = Suitability index

In the present study, all the thematic layers were integrated in IDIRISI GIS platform in order to prepare a map depicting suitable areas for artificial groundwater recharge. The total weights of each pixel of the final integrated layer were derived from the following equation;

$$S=(GM_w\ GM_{wf} + G_w\ G_{wf} + S_w\ S_{wf} + L_w\ L_{wf} + SL_w\ SL_{wf} + LD_w\ LD_{wf} + WF_w\ W_{wf}F + DB_w + DB_{wf}) \quad \ldots (2)$$

Where, 'w' represents the weight of each factor, while 'wf' represents the weight of the each class of the individual factor namely geomorphology (GM), geology (G), soil (S), landuse/cover (L), slope (SL), lineament density (LS), water table fluctuation (WF), depth to bed rock (DB). Thus the artificial groundwater recharge index 'S', which is dimensionless quantity that helps in indexing the probable ground water recharge zones in the area, was estimated using equation (2) for each pixel in the final integration layer and was regrouped into different classes with equal class interval to divide the entire study area into different artificial recharge zones. In the similar studies at West Bengal (Choudary *et al.*, 2006) used the same procedure.

Model Evaluation

Accuracy of models is another parameter that is important for selecting optimum model. Therefore, we need a criterion for comparing the accuracy of output model results. There are several criterion in groundwater recharge site selection for validation of the models some of them are well yield data, geo electrical resistively data and the well log analysis. However, in this method the validity of the model developed was checked against the bore well yield data which reflects the actual groundwater potential. Then by comparing the location status value and output maps estimated status, each correct estimating was identified by " +" and each incorrect estimating was identified by "-".

Results and Discussion

Geology

The study area comprises mainly of basalt and other trap intrusions. Detailed analysis revealed that there are six types of geological classes found, viz., laterites, sparse to moderate prophyrtic, sparse to moderate porphyrotic (compressed), high to moderate prophyrtic, mega cryst and moderate to high prophyrtic. The geological formations like laterite is considered better than the other geological formations for the recharge occurrence.

Soils

The major part of the study area is covered by vertisols (black cotton soil) and remaining part is with loamy soils (Fig. 3). There are, however, variations in the type of soil and different varieties of soil viz., very fine montmorillonitic (chromosters), fine montmorillonitic, loamy mixed lithic and very fine montmorillonitic. Recharge point of view, loamy mixed lithic soil is considered better than the montmorillonitic soil group.

Geo Morphology

The region has gently sloping topography. The major geomorphic land forms demarcated are pedimented plains, buried pediments, valley fills and residual /

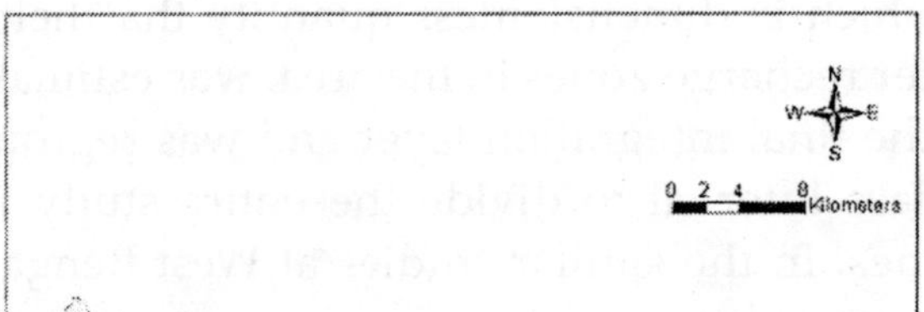

Fig. 3 . Soil map of the study area

hillocks (Fig. 4). Further, on recharge point of view, valley fill, buried pediment and pediment plains are favourable, while residual hills are least favourable.

Land Use / Land Cover

There are totally eight different land use/cover types were demarcated and shown in Fig. 5. The separability analysis and accuracy assessment was also carried out. The overall assessment of about 85 per cent was observed. In view of the recharge, cultivable lands, orchard and summer crops were given higher weightage whereas, settlement, wastelands were given low weightage.

Slope

The major part of the area is having gentle slope. However, variations in the slope (0 to 7.31 per cent) was made to group the entire area into four slope classes i.e., 0-1, 1-3, 3-5 and > 5 per cent. The areas having slope 0-1 and 1-3 per cent were assigned higher values and the areas, which are having slope more than 3 per cent were considered as the moderate in view of the ground water recharge.

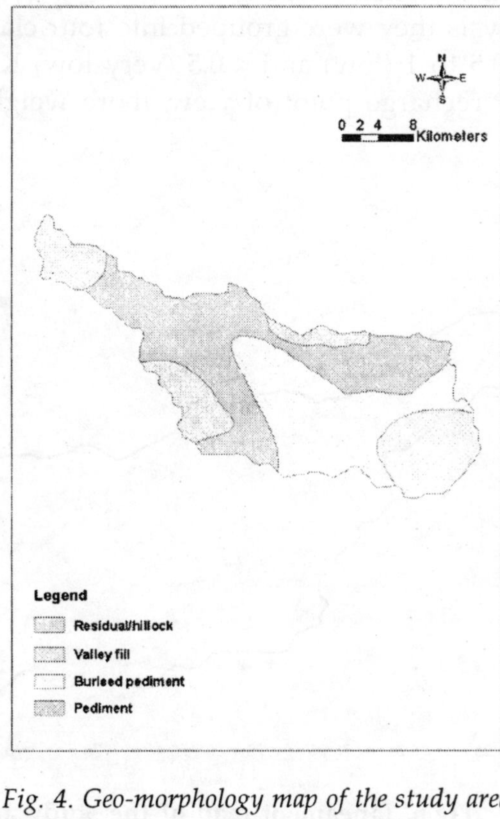

Fig. 4. Geo-morphology map of the study area

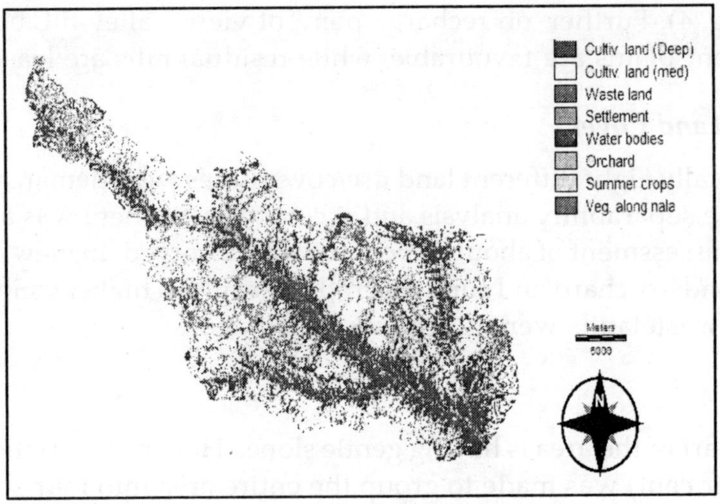

Fig. 5. Land use map of the study area

Lineament Density Map

In the study area, the lineament density was in the range of 0 to 2.96 Km/sq.Km. Hence, for the analysis they were grouped into four classes viz., > 1.5 (high), 1 to 1.5 (moderate), 0.5 to 1 (low) and < 0.5 (very low) Km/sq.Km respectively (Fig. 6). Further, for recharge point of view, more weightage was given where

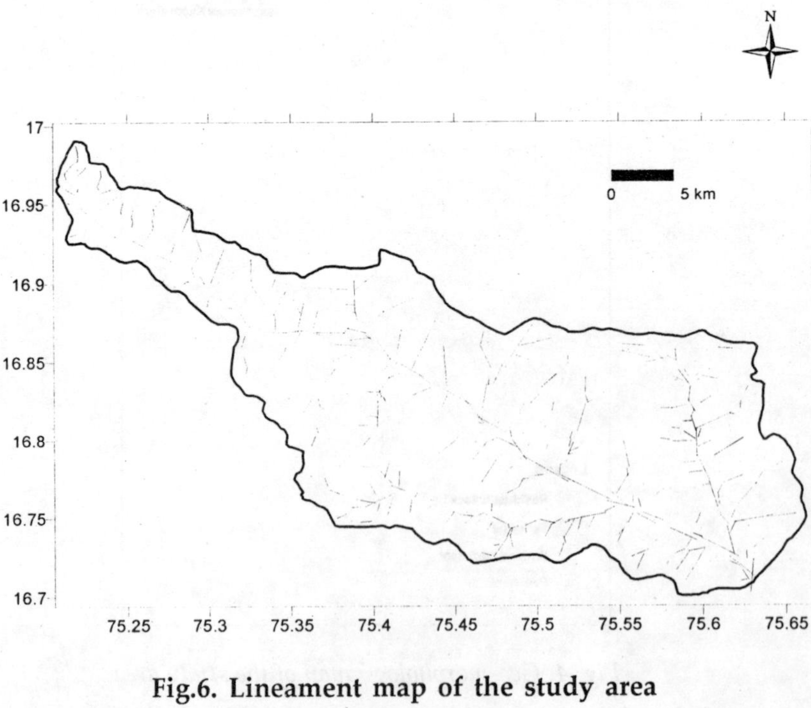

Fig.6. Lineament map of the study area

lineament density was higher than 1.5 Km/sq.Km, whereas, low weightage was given where the lineament density was lower than 0.5 Km/sq.Km.

Drainage Density Map

The dominant drainage pattern observed is dendritic to sub-dendritic. A glance at drainage density map reveals that density values ranged from 0 to 6.32 Km/Sq.Km. For the analysis purpose, they were re-grouped into four categories, i.e., high (> 3.5), medium (2-3), low (1-2) and very low (0-1) Km/Sq.Km. Considering recharge point of view, more weightage was assigned to low drainage density regions, whereas low weightage assigned to high drainage density regions.

Watertable Fluctuation

The fluctuation was in the range of 0 to 13.50 m. Except, small patches (< 3.78 per cent of the total area) on the lower and northern side of the watershed, the fluctuation was in the range of 0 to 6 m only, while, in the patches mentioned above, it was up to 13.5 m. For selecting artificial recharge sites, the maximum weightage was assigned to the regions with water table fluctuation more than 4 m, whereas less weightage was given where fluctuation was less than 4 m.

Depth to Bed Rock Map

The depth to bed rock was in the range of 15 to 77.78 m. For the analysis purpose, the depth to bed rock was re grouped into four classes viz., < 15 m, 15 – 30 m, 30 – 45 m and > 45 m. For assigning weightage for ground water recharge site selection, the regions with depth to bed rock more than 30 m was given high weightage, whereas, for the less than 30 m depth given the low weightage.

Data Integration for Suitability Analysis

The final weights of the polygons in the final integrated layer were obtained using equation 2 which is considered as the groundwater recharge index of individual polygons. The total range of the weights (150 to 300) of the polygons were classified into four classes viz., < 175, 175- 225, 225- 275 and > 275 by dividing the total class range by three. The weights of the polygons coming under each range were clubbed to divide the area into four different classes/zones viz., poor, moderate, good and excellent. The artificial recharge map thus obtained by the WLC method is shown in Fig. 7. The areal extent of the different suitability classes are fall in the good zone, followed by moderate (48%), poor (0.93%) and excellent (0.57%) zones. A close look at Fig. 7 reveals that the excellent zone in the form of small patches, spread over the south-eastern part of the study

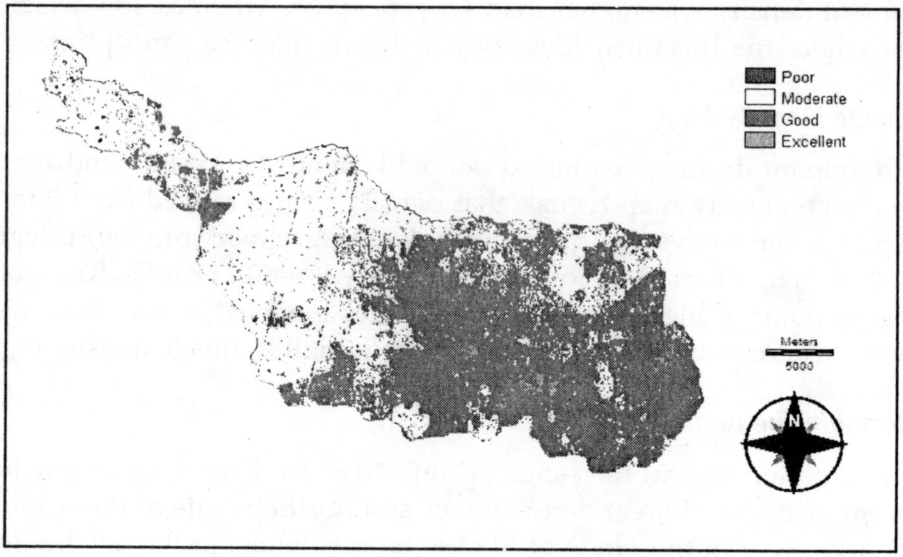

Fig. 7. Groundwater recharge site suitability map developed as per Weighted Linear Combination model.

area. The area was occupied by valley fill and buried pediments land forms and also the area is with low slope, high WT fluctuation and thick weathered mantle.

The good zone occupied most of the lower and small patches in the upper parts of the watershed. This area was occupied by buried pediment land form with low to moderate slope (0-1 and 1-3%), moderate WT fluctation (4-6m) and moderate to high aquifer thickness. The moderate zone was occupied by most of the upper and small patches in the lower part of the watershed. This area was covered by pediment and parts of residual hill type land forms, moderate to high slope areas and WT fluctuation was also relatively low. Further, the poor zone was found on the south western side and upper part of the watershed. This area was occupied by residual hillock land forms, slope is relatively high (> 5%) and WT fluctuation was low (2 to 4m). Further, the sub watershed wise various suitability classes are presented in the Table 1. A close look at data reveals that the most of the sub watersheds were dominated by either good or moderate suitability zones.

Evaluation of Models

The validity of the models developed was checked against the bore well yield data, which reflects the actual groundwater recharge potential. Yield of bore well ranged from 68.4 to 211.2 lpm. For the evaluation, the yield is in the range of > 200 lpm was put under excellent zone, 140 to 200 lpm in good zone, 80 to

Table 1
Sub watershed wise various recharge suitability classes as per weighted linear combination method

Sub-watershed No.	Suitability Classes				Total (sq. km)
	Poor (sq. km)	Moderate (sq. km)	Good (sq. km)	Excellent (sq. km)	
1	0.79	34.69	1.43	0.00	36.90
2	0.54	53.14	14.59	0.00	68.27
3	2.17	32.82	0.21	0.00	35.20
4	0.79	34.74	0.61	0.00	36.15
5	0.36	33.96	50.91	0.46	85.70
6	1.68	65.40	37.58	0.07	104.67
7	0.00	14.57	86.23	1.42	102.22
8	0.00	3.40	40.54	1.01	45.19
9	0.04	33.43	82.24	0.69	116.20
10	0.02	0.85	12.10	0.02	13.00
Total	6.39	307.00	326.44	3.68	643.50

140 in moderate zone and less than 80 lpm under poor zone. Then, by comparing location status value (well yield ranges) and output maps estimated status, each correct estimate was identified by "+" and each incorrect estimate was identified by "_" (Table 2). Table 2 shows the evaluation results for models accuracy. It can be seen from the Table 3 that the accuracy of the artificial recharge site selection by weighted linear combination (WLC) was 77.88 per cent. This is the fairly good estimation, therefore, it is proposed that for the study area, WLC model is an optimum model for artificial recharge site selection.

Conclusion

Satellite remote sensing IRS ID LISS-III digital data can be used successfully to classify the different land use/land covers, and Shuttle Radar Topographic Mission (SRTM) digital elevation model (DEM) data can be used to draw the classified slope maps. Multi Criteria Evaluation (MCE) based integrated models such as weighted linear combination (WLC) can be used successfully to locate suitable sites for the artificial recharge.

Table 2
Comparison between water yield data from field and prospective zones derived from the model

Sl.No	Status Location		Weighted Linear Combination	
	Village	Status	Status	Evaluation
1	Ratnapur	G	G	"+"
2	Ratnapur	G	G	"+"
3	Hosatti	G	G	"+"
4	Hosatti	G	G	"+"
5	Atalatti	G	M	"-"
6	Dashyal	E	E	"+"
7	Dashyal	E	E	"+"
8	Honwad	M	M	"+"
9	Honwad	M	G	"-"
10	Honwad	M	M	"+"
11	Bijjargi	M	M	"+"
12	Bijjargi	M	P	"-"
13	Rampur	M	M	"+"
14	Rampur	M	G	"-"
15	Tikota	M	M	"+"
16	Bannur	P	P	"+"
17	Bannur	P	P	"+"
18	Sarvad	G	G	"+"
19	kakamari	M	G	"-"
20	kakamari	M	M	"+"
21	kanmadi	M	M	"+"
22	Danyal	G	G	"+"

Excellent(E) > 200lpm

Good (G) = 140 to 200 lpm

Moderate(M) = 80 to 140 lpm

Poor (P) < 80 lpm

References

Bonham Carter and G.F., 1991. Geographic Information System for Geoscientists: Modelling with GIS. Pergamon, Ontario, pp.319-470.

Carver, S.J. 1991. Integrating multicriteria evaluation with Geographical Information Systems. *International Journal of Geographical Information Systems,* 5: 321-339.

Chowdhury, A., Madan, K. Jha, V.M. Chowdary, and B.C. Mal. 2006. Selection of artificial recharge zones in West Midnapur district of West Bengal, India: A GIS approach. *In*: An International Prospective on Environmental & Water Resources, organized by Environmental and Water Resources Institute of ASCE, New Dehli, pp.1-8.

Eastman, R. J. 2001. IDIRISI Andes; Guide to GIS and Image Processing., Clark University, USA., pp. 144.

Sahu, B.K. 2001. Groundwater Modeling in hard – rock terrain. *In*: Proceedings of national seminar on groundwater resources, MLS University, Udipur, India, pp.16-21.

Saraf, A.K. and P.R. Choudhury. 1998. Integrated remote sensing and GIS for groundwater exploration and identification of artificial recharge sites. *International Journal of Remote sensing,* 19(10):1825-1841.

6

Eco-friendly Management of Pod Borer, *Helicoverpa armigera* (Hubner) by Encouraging Predatory Birds and Other Bioagents in Chickpea Ecosystem

J.B. Gopali, S. Yelshetty, R.G. Teggelli, P.S. Dharmaraj and B.V. Patil

Introduction

India is bestowed with rich heritage of avian diversity consisting of 85 per cent of partially or fully insectivorous birds in different agroclimatic area (Anon., 1990; Salim Ali and Dillon, 1983; and Dhindsa and Saini, 1994). Several species of insectivorous birds have been found to feed on insect pests of crops including *H. armigera* (Chakravarty, 1988; Singh et al., 1990a, 1990b and 1990c) which are known to reduce the larval population to the extent of 84 per cent in Punjab. As enemies of insects, birds stand supreme among vertebrates (Sweetman, 1958). Among the predatory birds, black drongo *[Dicrurus adsimilis* (Bechstein)], house sparrows *(Passer domesticus* Linn.), blue jays *(Coracias bengalensis* Linn.), cattle egret *[Bulbulcus ibis* (Boddaotll, rosy pastor *(Sturnus rose us* Linn): and mynah, *Acridotheres ginginianus* Lathum) have been commonly recorded to be predators on large number of *Helicoverpa armigera* and lepidopteran insects on chickpea, pigeonpea, and groundnut crops (Beri *et al.*, 1972; Ghode *et al.*, 1988; Anon., 1987;Gokhale and Ameta, 1991 ; Yadava, 1991). The population of cattle egret and river stern *Sterna aurantia* was observed particularly more towards the end of January month (Ghode *et al.*, 1988). In addition, green bee eater, *Merops orientetis* Latham, grey strike and starling birds have been recorded as efficient predators (Anon.,1992a). It was also documented that the activity of predatory birds was found to be high when some grain crops are grown as intercrop in pigeonpea (Anon., 1992a ; Venugopal Rao, 1992). In Gujarat, about 300 rosy pastors *(S. rosens)* a migratory bird feed on *Achaea janata* for 8 days. Due to heavy predation, the number of large sized larvae and pupae are reduced. But the biotic agents alone are unable to control the outbreak of the pest (Parasharya *et al.,* 1988). Though India is bestowed with rich heritage of avian diversity (Salim Ali and Dillon, 1983 ; Dhindsa and Saini, 1994), beneficial role of insectivorous birds in the insect pest management has not received much recognition beyond faunisitic documentation. This is mainly because of over dominance of broad spectrum insecticides in the plant protection scenario (Gopali, 1998). Importance of birds faded further due to lack of awareness among the farmers about the behavioural manipulation of the predatory birds in pest management. In Karnataka, Gulbarga region has become a hot spot for *H. armigera* on chickpea. The encouragement of predatory birds has a special significance in chickpea ecosystem that leads to a harmonious and sustainable management of *Helicoverpa armigera* (Gopali *et al*, 2007 and 2008). Therefore, the development of implementation of ecofriendly management of chickpea pod borer is of paramount importance in achieving the sustainable production of chickpea.

Material and Methods

Standardization of Predatory Zones of Different Bird Perches

To make an assessment on the effect of bird perches on predatory activity of birds in chickpea (cultivar: Annigeri) ecosystem, a field trial was laid out in an Exploded Block Design (Muthuswami,1993) during 2003-04. Profused tree branches, nest type and T-shaped bamboo sticks as inanimate perch and sunflower and sorghum with goose neck as animate perch were evaluated at different predatory zones from 2 to 12 meter radius (Table 1). Inanimate perches of size four meters in length were erected at different distances from 2 to 12 meter radius. Similarly, animate perches such as sunflower and sorghum seeds were dibbled at different distances from 2 to 12 meter radius at the time of sowing of chickpea crop. During 2004-05, effective predatory zones of different bird perches were evaluated for their preference to alight on perches and consequential effect on the suppression of *H. armigera.* The observations on larval population and natural enemies such as predatory wasps, coccinellids and techinids were recorded at 50, 55, 60, 65, 70, 75 and 80 days after sowing (DAS). To record observation, one meter row length was selected randomly at two, four, six, eight, ten and twelve meters distances of each perch. Yield from one square meter area was recorded randomly from five spots for each treatment. The data were subjected to ANOVA and DMRT test to measure the efficacy of perch and effective zone of predation. Further, trials were laid out in a block design with an area of 2000

m^2 for each treatment where each block is divided in to four parts to evaluate the performance of bird perches at effective predatory zones on H armigera during 2005-06. The chickpea (cultivar- Annigeri-1) was sown at a spacing of 30 x 10 cm in the months of June - July for all the four years from 2004 to 2007 to carryout the bird experiments. The treatments as detailed under each study (Table 1 to 6) were implemented to compare the efficacy of treatments in comparison with insecticide as well as untreated check in all the field experiments. The observation on natural enemy population on live bird perches were recorded periodically. An experiment on effect of sunflower at different density on chickpea yield was carried out to know the effect on chickpea yield.

Assessment of Larval Population, Pod Damage and Grain Yield

Larval and natural enemy population were observed at 50, 55, 60, 65, 70, 75 and 80 days after sowing on one meter row length at five locations in each block. Pod damage was confirmed on the basis of large holes with regular margins and fully eaten seeds by *Helicoverpa*. It was estimated by counting the total number of pods and affected ones on ten randomly selected plants in each treatment. At

Table 1
Standardization of predatory zone of different bird perches in chickpea ecosystem during 2004-05

Predatory Zone (in radius)		Mean Larval Population Per Meter Row				
		Inanimate Perch			Animate Perch	
		Tree Branch	Nest Type with Bamboo Stick	T-shape with Bamboo Stick	Sorghum	Sunflower
1	2 m	0.71	0.65	0.39	1.96	0.35
2	4 m	0.82	0.80	0.81	2.02	0.86
3	6 m	0.62	1.41	2.20	3.18	1.94
4	8 m	1.89	2.57	3.59	3.90	3.76
5	10 m	2.13	2.82	4.24	4.47	4.23
6	12 m	2.39	3.04	4.39	4.42	4.44
7	No perch	4.23	4.72	4.60	4.84	4.90
CD at 5%		**0.33**	**0.37**	**0.45**	**0.41**	**0.48**

Observation on larval population recorded from 50th day onwards till 90th day at 5 days interval.

harvest, the pods from individual blocks were made in to three partitions and threshed separately to record yield data. Further, the yield data were computed to quintals per ha and subjected to statistical analysis. The details of the field experiments are given in the following headings. The data from two years were pooled and subjected to analysis of variance and means separated by DMRT.

Table 2

Predatory efficiency of different bird species in chickpea ecosystem during 2004-05

	Bird Species	Av Birds Visited Day/ac	Time spent /day (h)			No. of Larvae Consumed/ h / sq m²			Predatory Efficiency (%)
			Min	Max	Mean	Min	Max	Mean	
1	Mynah	84	1.5	3.20	2.35	3.4	7.2	4.5	35.3
2	Black drongo	7	1.2	3.25	2.23	2.2	3.4	2.9	25.2
3	Sparrows	58	1.7	2.30	2.00	0.7	2.3	1.5	10.8
4	Baya weaver	51	0.5	3.50	2.00	1.2	1.4	2.1	15.1
5	Cattle egret	5	0.5	3.20	1.85	0.5	1.4	1.9	13.7
6	Blue jay	2	0.1	0.50	0.30	NA	NA	NA	NA
7	Yellow sparrow	11	0.3	2.20	1.25	NA	NA	NA	NA
8	Green bee eater	4	0.2	0.50	0.35	NA	NA	NA	NA
9	Red walled lap wing	2	0.1	0.40	0.25	NA	NA	NA	NA
10	Hoopoe	2	0.1	0.30	0.20	NA	NA	NA	NA
11	King fisher	1	0.1	0.20	0.10	NA	NA	NA	NA

Sunflower used as live bird perch

Observation on larval population recorded from 50th day onwards till 90th day at 5 days interval

Considering the present costs of various inputs of all the treatments and the yield of pigeon pea, the C : B ratio values for each treatment are presented in Tables 5 and 6.

Results and Discussion

Encouragement of predatory birds and other bio-agents: Activity of birds from vegetative stage to grain filling stage was highest in chickpea ecosystem but decreased thereafter till the maturity of the crop. Peak activity of birds and other natural enemies were noticed from pod formation to grain filling stage when the crop was heavily infested by the pod borer. Among the several predatory birds, the most commonly noticed in chickpea ecosystem in Gulbarga were mynah, black drongo, egrets, variety of sparrows, green bee eater, blue jays and other small and medium omnivorous birds (Table 3). Therefore, the study of modern trends in utilizing the predatory birds for effective pest management strategies becomes inevitable. The detailed observation on behaviour of insectivorous birds in chickpea ecosystem is described in the Table-1 to 6.

Evaluation of Bird Perches

The results (Table-1) indicate that providing bird perches in chickpea reduced the pod borer density. Both inanimate perches (Profused tree branches, nest type and T-shaped bamboo sticks) as well as animate perches (sunflower and sorghum plants) provided alighting place from where birds have closer examination for the presence of prey. Among the inanimate perches, profused tree branch was more effective than T-shaped bamboo sticks, while sunflower was more efficient than sorghum plants among animate perches. While the perched area at two meter radius with inanimate and animate perches recorded 0.39 to 0.71 and 0.35 to 1.96 larvae per meter row respectively, unperched area registered 4.23 to 4.90 larvae per meter row (Table 1). The animate perches such as sunflower and sorghum perches found to be as effective as inanimate perches at four meter radius suggesting easy adoptability and acceptability of live bird perches for managing the pod borer. The variation in effective radius of each perch suggest that number of perches per hactare would vary according to the type.

Sunflower is fast growing plant which acted as live bird perch from vegetative stage to till harvesting and checked the larval population below ETL (Economic Threshold Level) throughout the cropping period of the chickpea whereas sorghum being slow in growth acted as bird perch and checked the larval population after the emergence of ear head. Interestingly, sunflower not only acted as live bird perch but also acted as trap crop for *Helicoverpa* larvae. Field evaluation of different perches with varying predatory zones revealed that mere

Table 3

List insectivorous birds recorded in chickpea ecosystem

Sl. No	Common Name	Scientific Name	Family	Order	Predatory Behaviour	Type of Visit	Feeding Habit
1.	Mynah	*Acridotheres tristis*	Saturnidae	Passeriformes	Gound predator	Regular visitor & gregarious (50 to 250)	Omnivorous
2.	Black drongo	*Dicrurus adsimilis*	Dicruridae	Passeriformes	Aerial predator	Regular visitor & Solitary	Carnivorous
3.	House sparrow	*Passar domesticus*	Ploceidae	Passeriformes	Ground predator	Frequent visitor & gregarious (10 to 100)	Omnivorous
4.	Baya weaver	*Ploceus philippinus*	Ploceidae	Passeriformes	Aerial and ground predator	Frequent visitor in 2 to 5 numbers	Omnivorous
5.	Common babbler	Turdoides caudatus	Timaliidae	Passeriformes	Aerial predator	Regular visitor & Solitary	Omnivorous
6.	Blue jay	*Coracias benghalensis*	Coracidae	Coraciformes	Aerial predator	Occasional visitor, shy bird & Solitary	Carnivorous
7.	Cattle egret	*Bulbulcus ibis*	Ardeidae	Ciconiformis	Ground predator	Frequent visitor in Small groups (5 to 50)	Carnivorous
8.	Green bee eater	*Merops orientalis*	Meropidae	Coraciformes	Aerial predator	Occasional visitor in pairs	Carnivorous
9.	Red watted lap wing	*Vanellus indicus*	Charadriidae	Charadriiformes	Ground predator	Occasional visitor in 2 to3 in numbers	Omnivorous
10.	Hoopoe	*Upupa epops*	*Upupidae*	*Coraciformes*	Ground predator	Occasional visitor & solitary	Omnivorous
11.	Swift tail flyer	*Apus affinis*	Apodidae	Apodiformes	Aerial and ground predator	occasional visitor & gregarious (25 to 100)	Omnivorous
12	Swallows	*Hirundo rustica*	*Hirundinidae*	*Passeriformes*	Aerial and ground predator	Occasional visitor & gregarious (20 to 50)	Omnivorous

100 seed weight of sunflower = 5.20 g, For 625 seeds = 32 grams = 50 grams/ha

100 seed weight of sunflower = 3.20 g, For 625 seeds = 20 grams = 50 grams/ha

Table 4

Performance of bird perches at effective predatory zones on *H. armigera* in chickpea during 2005-06

Sl. No.	Type of Bird Perch	Predatory Zone in (Sq m)	Mean Larval Population of *H. armigera* / m row 50 DAS	55 DAS	60 DAS	65 DAS	70 DAS	75 DAS	80 DAS	Mean	Yield (g)/ sq.met	Yield (q/ha)	No. of perch/ ha	Perch : crop Ratio
1	Tree branch	10	0.44	0.72	1.77	2.5	2.58	1.65	0.6	**1.47**	51.2	524	100	2.35625
2	Nest type	8	0.28	0.56	1.85	2.8	2.7	1.62	0.56	**1.48**	49.04	496	100	1.5847222
3	T- perch	6	0.36	0.66	1.96	2.6	2.4	1.42	0.54	**1.42**	55.4	574	150	1.5847222
4	Sorghum	4	0.84	1.44	2.82	3.84	3.3	2.36	0.65	**2.18**	48.4	489	625	0.4118056
5	Sunflower	4	0.32	0.64	0.86	0.75	0.88	0.54	0.44	**0.63**	79.4	894	625	0.4118056
6	Profenphos	-	0.12	0.28	1.24	0.48	0.44	0.64	0.4	**0.51**	98.8	988	-	-
7	No perch	-	1.66	1.4	2.9	4.4	4.72	2.28	1.14	**2.64**	39.4	394	–	–
	CD at 5%		**0.25**	**0.29**	**0.33**	**0.36**	**0.41**	**0.32**	**0.27**	**0.29**	**7.24**	**94.8**		
	SEm ±		**0.08**	**0.09**	**0.11**	**0.12**	**0.13**	**0.1**	**0.09**	**0.09**	**2.41**	**31.6**		
	CV (%)		**9.48**	**10.25**	**11.85**	**15.65**	**12.4**	**9.6**	**17.2**	**7.8**	**13.6**	**11.64**		

100 seed weight of sunflower = 5.20 g, For 625 seeds = 32 grams = 50 grams/ha

100 seed weight of sunflower = 3.20 g, For 625 seeds = 20 grams = 50 grams/ha

Table 5

Comparative performance of effective bird perches in chickpea ecosystem during 2006-07

Sl. No.	Type of bird perch	No. of Perches/ha	Mean larval population / m row					Yield (q/ha)	Gross Income	Input cost	other Cost	Total cost	Net Returns	B:C Ratio
			50 DAS	60 DAS	70 DAS	80 DAS	Mean							
1	Sorghum	50 grams	2.14	2.80	3.54	1.44	2.48	6.20	15500	5	3175	3180	12320	4.87
2	Sunflower	50 grams	0.80	0.55	0.35	0.38	0.52	8.75	21875	10	3175	3185	18690	6.87
3	T – perch	150 No	1.66	1.92	2.58	1.65	1.95	7.38	18450	200	3175	3375	15075	5.47
4	Sorghum + T-perch	50 g + 75 n	1.59	1.58	1.87	1.42	1.62	7.52	18800	205	3175	3380	15420	5.56
5	Sunflower + T-perch	50 g + 75 n	0.62	0.55	0.84	0.40	0.60	9.16	22900	210	3175	3385	19515	6.77
6	Sorghum+Sunflower	50 g+ 50 g	0.60	0.47	0.41	0.36	0.46	9.64	24100	15	3175	3190	20910	7.55
7	Sorghum+Sunflower+T-Perch	50g+50g +50 n	0.51	0.35	0.30	0.25	0.35	9.72	24300	220	3175	3395	20905	7.16
8	Profenphos 50EC (3 ml/lit)	----	0.17	0.21	0.13	0.16	0.15	9.94	24850	1700	3175	4875	19975	5.10
9	UTC (No perch)	----	2.26	3.14	4.67	2.97	3.26	5.12	12800	0	3175	3175	9625	4.03
	CD at 5%		0.27	0.31	0.34	0.21	0.25	105.4						
	SEm ±		0.09	0.10	0.11	0.07	0.08	34.8						
	CV (%)		7.41	9.50	10.54	14.05	11.33	24.60						

Current rate of chickpea : Rs 2400/q

Other cost : Includes cost of cultivation according to the package of practice

Table 6

Large scale evaluation of live bird perches in comparison with insecticide at different locations of North Karnataka during 2006-07 and 2007-08

Sl. No	Location	Area in ha	T1*				T2**	% Increase in Yield Over T2
			Chickpea Yield (q/ha)	Sunflower Yield (q/ha)	Equivalent Chickpea Yield (q/ha)	Total Chickpea Yield (q/ha)	Chickpea Yield (q/ha)	
Large Scale Demonstration during 2006-07								
1	Agricultural Research Station, Gulbarga	0.40	8.92	0.88	1.02	9.94	9.14	8.76
2	Kalagi, (Tq-chittapur :Dist- Gulbarga	2.00	9.66	1.30	1.52	11.18	10.17	9.90
Large Scale Demonstration during 2007-08								
3	Kabirbad wadi (Tq-Humnabad:Dist-Bidar)	0.40	9.84	1.40	1.63	11.47	10.25	11.93
4	Tadola (Tq-Aland : Dist-Gulbarga)	0.40	11.72	1.25	1.46	13.18	12.2	8.02
5	Purthipalli (Dist- Raichur)	0.40	9.47	1.08	1.26	10.73	9.95	7.84
6	Agricultural Research Station, Malnoor	0.40	8.23	1.20	1.40	9.63	8.56	12.50
7	Agricultural Research station,Gulbarga	0.80	9.90	1.10	1.28	11.18	9.44	18.47
8	Gundagurti (Tq-Chittapur: Dist-Gulbarga)	2.00	12.24	0.95	1.11	13.35	12.72	4.94
9	Hagarga (Dist- Gulbarga)	1.60	11.94	1.25	1.46	13.40	11.25	19.10
	Mean		**10.21**	**1.16**	**1.35**	**11.56**	**10.41**	**11.27**
A	**Gross income*** **					27749	24981	
B	**Cost of inputs**					15	1700	
C	**Other cost********					3175	3175	
D	**Total cost (B+C)**					3190	4875	
E	**Net returns (A-D)**					24559	20106	
F	**B:C ratio (A / D)**					**8.70**	**5.12**	

* T1 – Sowing of sunflower and sorghum seeds @ 50 g/ha each along with chickpea seeds at the time of sowing

** T2 –Two sprays of Profenphos 50EC @ 3 ml/l one at 50% flowering and other at pod formation stage

type of perch is not enough to prove its acceptability. The number of perches is also equally responsible for increased efficiency of perch (Table 4). Considering the effective predatory zone, economic and operational feasibility and adoptability of the perch, the number of perches required for one hactare was computed for each type of perch. While this number reached to 625 ha' for animate (sunflower and sorghum) perches and it ranged from 100 to 150 for inanimate (tree branches and T-shaped bamboo sticks) perches (Table 4). Again, among the former category, sunflower was more efficient as compared to sorghum plants. It is mainly because of fast growth, rigid support and increased height of the perch over the crop canopy right from vegetative stage till pod maturity stage of chickpea. Since drongo is a solitary and aerial predator, it utilized each and every sunflower plant efficiently without competing for alighting place. Based on seed weight and number required per ha, about 50 grams of sunflower and 50 grams of sorghum seeds per hectare are required to sow along with the chickpea crop (Table 4). This treatment not only found cost competitive but also ecologically balanced and economically viable as compared to other perch combination. It has been documented that the activity of birds is found high when some grain crops are grown in pigeonpea as intercrops (Anon., 1992b; and Venugopal Rao *et aI.*, 1992 ; Gopali, 2007 and 2008). Similarly, many authors highlighted the importance of insectivorous birds which need to be manipulated in the agroecosystem using appropriate management practices (Chakravarthy, 1988 ; Parasharya *et al.*, 1988 and 1994; Gopali, 1998).

Ground Predators

Bird species that come in flocks used perches as alighting places to rest for a while and scanned for availability of host. Among the ground predators mynah, sparrows and baya were more efficient in predating the chickpea pod borer, *Helicoverpa armigera*. House sparrows, yellow sparrows and baya, in flocks (10-15), hovered at the basal branches of chickpea plants in search of the prey at the base of the plants and picked the larvae in their beaks contributing 26% of larval reduction. Similarly, mynah birds, in flocks (50-200) contributed 35% reduction in larval reduction. These birds also preyed most of the larvae that were entering into pupation in soil (Beri *et al.*, 1972 ; Patel *et al.*, 1987 ; Singh *et al.*, 1990a, 1990b and 1990c; Gokhale and Ameta, 1991; Gopali *et al.*, 2007). As these insectivorous birds exert greater pressure on the pest density in reasonably short time, it helps in the management of obnoxious pest like *H. armigera* in chickpea ecosystem.

Aerial Predators

Among the aerial predatory birds which landed on both animate and inanimate perches, black drongoes (25%), were the most experienced birds as they located and consumed more number of larvae in a shortest time. Hence, there was marked reduction in larval population up to 4 meter radius from the perches. The larvae on pods measuring between 1.5 to 2.5 cm were easily detected and picked by black drongo even from four meters from animate perches. Late instar larvae were preferred (80%) than early instar (20%). On the basis of residual population of larvae in the predatory zone, animate perches proved to be more effective in reducing the damage and increasing the yield. Similar observations were made by Gopali *et al.* (2007 and 2008) on the utilization of birds through behavioural manipulation in the chickpea ecosystem. Differential rate of predation by the birds was reflected in the grain yield. While yield per square meter area was 79.40 grams in sunflower perched area, it was 39.40 grams per plant in unperched area (Table 4). The yield realised from each square meter decreased with increase in radius distance from perch. These figures convincingly demonstrate the superiority of sunflower perches over the other perches. These findings convincingly demonstrate the superiority of animate perches (sunflower and sorghum) which was obviously due to more number landing areas without competition for alighting places. These results corroborate with Gopali (1998) and Gopali *et al.*, (2007 and 2008) who opined that both sunflower and sorghum perches are more effective in chickpea ecosystem.

Nevertheless, unlike drongo, jays were less efficient as they were unable to recognise the green coloured, motionless and partly exposed larvae beyond two meters radius from perches. Therefore, it may be inferred that black drongo, due to its innate capacity to recognise and pull out the green and partly exposed caterpillars, proved to be the most effective aerial predatory bird in the chickpea ecosystem. Thus, black drongo dominate in the utilization of perch and is exclusively insectivorous, voracious and highly intelligent bird. These birds remain active even up to dusk in the field. Hence, this bird is regarded as an intimate friend of farmer and greatest foe of caterpillars in pigeon pea ecosystem. The observations on the predatory role played by drongo by Yadav (1991); (Anon.,1992a and 1992b) support the current findings. Voracious feeding habit, delectability of larvae and accessibility of the prey in chickpea ecosystem, even under high predatory risk were the attributes to make drongo as the potential predator. Secondly, the drongo has unique physique: deep forked tail with slim body facilitate graceful, amazing, effortless and agile turning and twisting movements for hunting the prey dexterously in the air. Soon after detection of the larvae, the prey is either caught under claws or picked by beak and settled on the perch. The larva is pound repeatedly till immobilization and torn into pieces to aid in swallowing. The information on behavioural manipulation of insectivorous birds is scanty. These observations suggest that type of vegetation, crop canopy, predatory habit, alighting support and agro-climatic conditions are possible factors to make change in the spectrum the bird species in a given ecosystem.

Predatory Efficiency

Based on regular arrival, time spent in field and number of larvae consumed, highest predatory activity was noticed in chickpea ecosystem by variety of sparrows (11%) baya's (15%), mynah (35%), black drongo (25%),which played a commendable role in reducing the pod borer population (Fig. 2). Among these birds, black drongo (1.20 to 3.25 hr/day), sparrows (1.70 to 2.30 hr/day) and mynah's (1.50 to 3.20 hr/day) spent longest time in the field and, devoured highest number of larvae (Table 2) as compared to rest of the birds. The observations indicate that the type of crop and pest density are important factors to influence the bird species and their predatory behaviour. These results corroborates with the observations made by Anon (1987) ; Ghode *et al.*(1988) and Gopali (1998). The detailed behavioural study on of predatory birds in chickpea ecosystem were given in the Table 6.

Encouragement of Other Bioagents

The basic principle involved in the study is that most of the *H. armigera* larvae existing on chickpea migrated to sunflower plant right from vegetative stage till pod filling stage of chickpea crop and each sunflower plant recorded 3.70 to 4.20

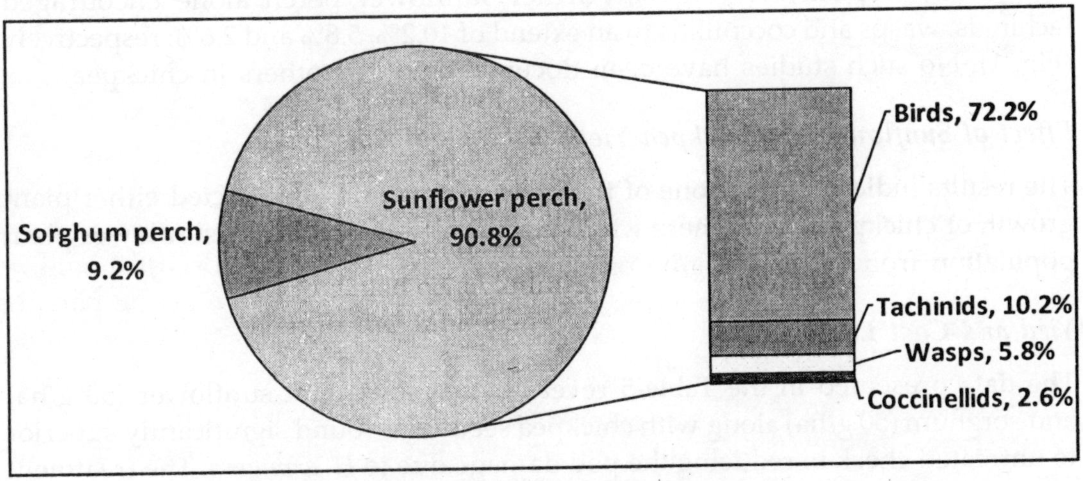

Fig 1 : Percent population of natural enemies due to live bird perches in chickpea ecosystem

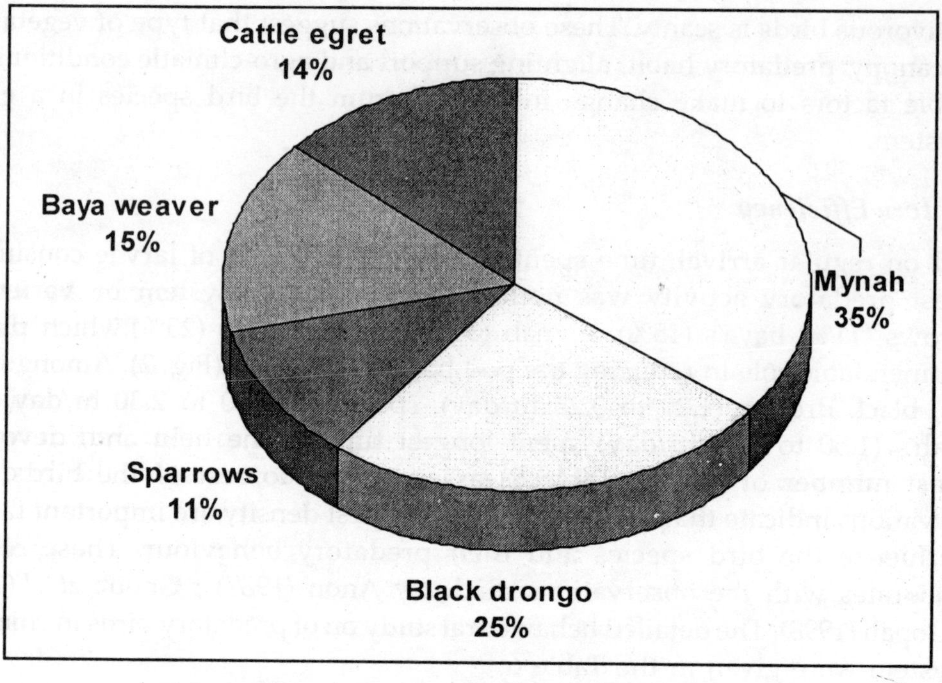

Fig 2 : Predatory efficiency of major insectivorous birds in chickpea ecosystem

per plant. Further, these larvae were easily detected by predatory birds as well as easily accessible to wasps, tachinids and coccinellids. The encouragement of natural enemy population due to sorghum and sunflower perches was 9.20% and 90.8%, respectively (Fig 1). Further sunflower perch alone encouraged tachinids, wasps and coccnillids to an extend of 10.2%, 5.8% and 2.6%, respectively (Fig. 1). No such studies have been documentation by others in chickpea.

Effect of Sunflower on Chickpea Yield

The results indicated that none of the density of sunflower affected either plant growth or chickpea yield. There was no significant difference of sunflower plant population from 50 to 150 g/ha on chickpea.

Yied and Cost Economics

The data presented in the Table-5 revealed that sowing of sunflower (50 g/ha) and sorghum (50 g/ha) along with chickpea seeds was found significantly superior to untreated check in reducing the pod damage due to *H. armigera*. The treatment recorded maximum grain yield of 9.64 q per ha which, however, was at par with insecticide indicating similarity in their performance. Apparently, all types of perches were significantly superior to untreated check which recorded lowest yield of 5.12 q/ha. Considering the present costs of various inputs used in all the treatments and the yield of chickpea, the B:C ratio for each treatment was worked out and the values are furnished in Table-5. It was concluded that highest net returns (Rs. 20910) and B:C ratio (1:7.55) with sunflower (50 g/ha) and sorghum (50 g/ha) when sown along with chickpea seeds.

Large Scale Validation of Live Bird Perches

The study was verified with large scale demonstration during 2006-07 and 2007-08 at various locations in farmer fields in Northern parts of Karnataka, India. The results indicated that sowing of sunflower and sorghum seeds @ 50 g/ha each along with chickpea seeds at the sowing time registered a chickpea yield of 10.21 q/ha with addition yiield of sunflower (1.16 q/ha) and sorghum (10-12 kg/ha) resulting in highest net returns (Rs. 24559/ha) and benefit to cost (1:8.70) ratio as compared to profenphos 50EC as standard check (1:5.12) suggesting low cost technology, eco-friendly, easily acceptable and adoptable by the farmers for managing the chickpea pod borer.

Conclusion

Use of synthetic chemicals particularly against gram pod borer, *H. armigera* in chickpea ecosystem is too expensive for the ordinary farmer which has led to development of a non-insecticidal approach by encouraging natural enemies such as predatory birds, coccinellids, wasps and tachinids in chickpea ecosystem. The study revealed that sowing of sunflower (50 g/ha) and sorghum (50 g/ha) along with chickpea seeds recorded lower larval number (0.35 larva/m row) over no bird perch (4.12 larva/m row). Sunflower acted as most suitable live bird perch in chickpea ecosystem as it is very fast growing plant and provided rigid support for alighting insectivorous birds right from vegetative stage till maturity of the crop. Major predatory birds attracted to sunflower perch were mynah (*Acridotheres ginginianus* Lathum), black drongo (*Dicrurus adsimilis* Bechstein), cattle egret (*Bulbulcus ibis* Boddaot), baya's and variety of sparrows which contributed to 40.20%, 20.80%, 12.40%, 10.5%, 7.20 % and 8.90 predation, respectively in reducing the larval number in the shortest time. Observation on predatory wasps carrying large number of larvae recorded on sunflower plant. The number of coccinellids (0.32/plant) were higher on sunflower plant. Similarly population of tachnid fly (0.57 fly/plant) was highest accounting 75 to 90% parasitization of late instar larvae. Thus, both sunflower and sorghum plants encouraged insectivorous birds and other bio-agents in chickpea ecosystem as a non-insecticidal approach for sustainable management of *H. armigera*.

References

Anonymous (1987) *Annual Report of All India Co-ordinated Research Project on Biological Control of Crop Pests and Weeds*. Indian Council of Agricultural Research IIHR, Bangalore, India.

Anonymous (1990) *Birds*. Publication and Information Directorate, CSIR, Hillside Road, New Delhi, India, pp.82.

Anonymous (1992a) *Annual Report of All India Co-ordinated Research Project on Biological Control,* NCIPM, Bangalore, India.

Anonymous(1992b) *Research Highlights of AICRP on Agricultural Ornithology*, APAU, Hyderabad, pp. 10-20.

Beri Y. P., Raizada M and Bhatnagar R.K. (1972) Food in adults of common sparrow, Passer domesticus Linn. *Entomologists Newsletter*, 10 : 65-66.

Campbell R.W and Sloan R.J. (1977) Natural regulation of innocuous gypsy moth population. *Environment Entomology*, 6 : 315-322.

Chakravarthy A.K. (1988) Bird predators of pod borers of field bean (*Lablab niger*). *Tropical Pest Management*, 34 : 395-398.

Dhindsa M.S. and Saini H.K. (1994) Agricultural ornithology : an Indian perspective. *Journal of Bioscience*, 19 : 391-402.

Ghode M. K., Nayak U.K., Ghosh P.K. and Pawar A.D. (1988) Avian predation of gram pod borer (*Helicoverpa armigera*) in Orissa. *Journal of Advanced Zoology*, 9 : 148-151.

Gokhale V.G. and Ameta O.P. (1991) Predatory behavious of house sparrow, *Passer domesticus* L. In the population regulation of *Heliothis* sp. Infesting chickpea, *Cicer arietinum. Indian Journal of Entomology*, 53 : 631-634.

Gopali J.B. (1998) Studies on integrated pest management of pigeonpea pod borer, *Helicoverpa armigera* (Hubner) with special reference to HaNPV and insectivorous birds. *PhD Thesis*, University of Agricultural Sciences, Dharwad, India, pp.185.

Gopali J.B., Suhas Yelshetty, Raju Teggelli and Mannur D.M. (2007) Eco-friendly management of pod borer, *Helicoverpa armigera* (Hubner) by encouraging predatory birds in chickpea ecosystem. Paper presented in *National Symposium on Legumes for Ecological Sustainability : Emerging Challenges and Opportunities*, November 3-5, held at Indian Institute of Pulses Research, Kanpur, p. 168.

Gopali J.B., Siddalingesh, Raju Teggelli and Suhas Yelshetty (2008) Live bird perches - a non-insecticidal approach for the management of chickpea pod borer, *Helicoverpa armigera*. Paper presented in *First International Conference on Agrochemicals Protecting Crop, Health and Natural Environment*, 8-11 January, held at IARI, New Delhi, p. 214.

Muthuswami M (1993) Development of tank-mix formulations of nuclear polyhedrosis virus against *Helicoverpa armigera* (Hubner). *PhD. Thesis*, Tamil Nadu Agriculture Unviersity, India. P. 100.

Parasharya B.M., Dodia J.F., Mathew K.L. and Yadav D.N. (1994) Natural regulation of white grubs (*Holotrichia* sp. Scarabidiae) by birds in agroecosystem. *Journal of Biosicience*, 19 : 381-389.

Parasharya B.M.,. Dodia J.F., Yadav D.N. and Patel R.C. (1988) Effect of bird predation and egg parasitism on castor semilopper, *Achaea janata* L. (Lepidoptera : Noctuidae) in Gujarat. *Journal of Biological Control*, 2 : 80-82.

Patel H.M., Yadav D.N., Parasharya B.M. and Patel R.C. (1987) A note on predation of *Catopisillia* sp. (Lepidoptera : Pieridae) by birds and wasp. *Journal of Bombay Natural History*, 84 : 452-454.

Salim Ali and Dillon R.S. (1983) A pictorial guide to the birds of the Indian Subcontinent. *Bombay Natural History*, Oxford, pp. 176-188.

Singh J., Sidhu A.S., Dhindsa M.S. and Saini H.K. (1990c) The effect of fodder cuttings on larval population, disease incidence and bird predation of *Helicoverpa armigera* in berseem. *Annual Biology*, 6 : 153-159.

Singh K.J., Sandhu S.S. and Singla M.L. (1990a) Ecology of *Helicoverpa armigera* (Hubner) in chickpea in Punjab. *Journal of Insect Science*, 3 : 47-52.

Singh T.V.K., Singh K.M. and Singh R.N. (1990b) Groundnut pest complex succession of pests. *Indian Journal of Entomology*, 52 : 493-498.

Sweetman H.L. (1958) The principles of biological control. Dubuqus, lowa ; W.C. Brown Co., pp. 56-64.

Torgersen T.R. and Campbell R.W. (1982) Some effects of avian predators on the western spruce bud worm in north central Washington. *Environmental Entomology*, 11 : 429-431.

Torgersen T.R. and Mason R.R. (1987) Predation of egg masses of the Douglas fir, Tussock moth (Lepidoptera : Lymantridae). *Environmental Entomology*, 16 : 90-93.

Venugopal Rao N., Tirumala Rao K., Sathyanarayana Reddy A. and Meera Gupta (1992) Control of pigeonpea pod borer, Helicoverpa armigera (Huber) with nuclear polyhedrosis virus. *Journal of Biological control*, 4 : 122-124.

Yadav C.P.S. (1991) White grub management in groundnut. Technical bulletin No.1, AICRP (White grub), Agricultural Research Station, Durgapur, Jaipur, pp. 14.

7

Sustainable Agricultural Development

K.A. Rasure

Introduction

Agriculture and environment have an age old and permanent relationship with each other. Perhaps no other human effort has interacted with environment as easily as agriculture did. Yet the concern for environmental degradation is of recent origion. This often creates the impression that agriculture is not in conflict with natural harmony and does not have adverse effects on ecology. The blame is often put on industry that emits huge amount of smoke and other pollutants. This may partly be true as it is not wholly responsible for the environmental problems that the society is faced with these days. Quite a lot has happened in agriculture in this century that has contributed its share of ecological disorders. Again, it is being felt that new agricultural technologies using an increased dose of inorganic intents have adversely affected nature, and that traditional agriculture has been neutral to it and its operations. This may again be a fallacy but it also points to the fact that with the increased adoption of inorganic technologies, conflicts of agriculture with natural equilibrium may have increased considerably. That mankind is closely linked with environment and its very existence is likely to be threatened if natural disorders go beyond a point, should sound be enough a warning to all of us to direct our keen attention to the problems of environment and the measures to be adopted to rectify the disorders that have already taken place and to prevent future disequilibrium taken place. It is in this respect that a study of agricultural development and environment has assumed importance. In the present national seminar, an effort will be made to examine the harmful effects of agricultural growth on environment and what can be done to prevent these effects taking place in future.

Environment & Sustainable Agricultural Development

Ever since man started realizing the existence of his entrepreneurial qualities and the urges of going his own way, the exploitation of nature began. When man no longer wanted to gather food that nature produced for him he started working against it by clearing patches of land and growing the food he wanted. Soon came in the felling of trees for fuel wood and then for timber leading to the clearance of forest areas, which later became responsible for the degradation of soil. Man started this onslaught against nature out of a compelling situation emanating from the multiplication of mankind. Had nature drawn its limits beyond which man's onslaught would not be permitted, perhaps the nature of things would have been entirely different. Nature has learnt only to give and not to refuse and the aggressive man never relented in his efforts to exploit the natural resources.

Agriculture had to be in the lead in bringing about the natural disorders as its ever increasing greed for land went on unabated over time the intensity of harm that agricultural growth has done to nature has increased owing to man's compulsion of producing more and more good and other agricultural commodities and in the process, using more and more inorganic inputs in farm operations which have come to be identified as harmful agents in nature's operation.

The technology developed over time, did not help much. While it enabled the deepening of the resource-use at its present level, it certainly did not prevent further exploitation of whatever was left. Much less effort was made to replenish the resources.

The impact of agricultural development on the environmental situation is now more complex and manifold. Agriculture is heavily dependent upon industry and many a produce produced by the industry and causing harm to environment are meant for agricultural development only. Thus, industry and agriculture together, are harming the environment in a greater measure that at any time in the past. New agricultural technology having many inorganic components has produced highly harmful effects on nature. It is reported to have affected the earth's crust and the useful bacteria, besides upsetting the ecological balances in its own way.

An increased use of chemical fertilizers has been hardening the upper layers of the soil and with each successive crop season, the tractors have to be so designed as would dig deeper and deeper. With each such exercise, the depth of the hard layer goes on increasing this makes the soil barren for future cultivation. For how long shall tractors be used to dig deeper should not be hard to understand and unless the process, is immediately reversed through some alternative technology, it might threaten the whole process of agriculture in not a too distant future.

Many a times, it has come to notice that during the rainy season, the highland soils subjected to inorganic inputs do not retain rainwater in any appreciable measure. The whole of the rainwater flows down. It acts as a double-edged weapon. The high lands do not retain moisture, which is essential for living organisms such as earthworms and other useful bacteria which in turn are highly conducive to the growth of agricultural crops and on the other hand, the low low-lying areas get inundated, threatening the standing crops and also bringing large-scale destruction to these areas.

Inorganic farming has also resulted in the crop genetics suffering from many diseases and new rusts have come to be known which delay all advances in agricultural research. New pests develop and crop-genes become resistant to new pesticides and other inorganic treatments. It is reported that food losses owing to pests and diseases are about one-third of the total production.

The foregoing discussions amply demonstrate how agriculture and environment can come in conflict with each other. That his conflict was not recognized in time has led to different forms of environmental degradation, like deforestation, erosion of soils, flood havocs, water logging, air and water pollution, besides other problems of soil destruction. The adverse effects of agriculture and its growth on environment may be more indirect that in industry. Sustainable agricultural development is needed ; therefore, an irrigation project has to be undertaken or a multipurpose project is to be taken up. These projects result in several problems such as destroying plant-life at one place and submerging vase forest areas at another, displacing people at one place and their rehabilitation elsewhere, clearing forests at one place and attempting to compensate for these at an alternative place. Can these displacements be perfectly substituted elsewhere is quite a difficult job. While such projects are immensely important for Sustainable Agricultural Development, how much cost do we have to pay in terms of degrading environmental, condition has to be carefully considered. It should not so happen that while safeguarding out present, we mortgage our future.

One of the important elements of the new farm technology is assured irrigation. In areas where irrigation was assured, it worked wonders including exploitation of avenues of irrigation in other areas. This resulted in digging of wells on a large-scale. In the process, instead of optimal use of water, excessive use was made.

This affected in two ways. On the one had, the problem of salinity arose and on the other, the water table started going deep down to an alarming proportion. This over-excitement of adopting the new techniques makes us indifferent towards the future consequences and hence excessive use of the revolutionary technology. In some places, even water-logging is reported to be causing concern. In a country of acute scarcity of land, wasting some areas owing to water-logging can be ill-afforded. The proportion of water-logging area to the total irrigated area is reported to be varying between 7 per cent to around 30 per cent. This only points to the dimensions of the danger involved in the development of agriculture when it is not farmed thoughtfully. Similar problems arise with respect to those activities where forest resources are used. Excessive use of already depleting forest resources without giving sufficient attention to its regeneration is causing concern.

Remedial Measures

Future course of agricultural development, therefore, shall have to chalked out carefully if it is to be ensured that no more degradation of our environment takes place. Some of the important points to be kept in mind may be listed as under.

1. Suitable water management.
2. Suitable afforestation policy.
3. Suitable input mix.
4. Suitable cropping pattern.
5. Suitable rural development strategies.

1. **Water Management** : Water is the life line of our fields and an important component of the new inorganic technologies and hence there always remains a stron temptation to over-use water. Not only water needs to be used rationally but proper awareness shall also have to be created against the evil effects of excessive water application. The distribution channels of limited water supplies shall have to be designed carefully so that wastage of water is avoided. Digging so many wells by individual farmers may neither be economical nor even desirable. Community wells may be dug and supply of water granted on need-base at uniform prices. Command areas should pay greater attention to water management, so that results are optimized. Salinity and water logging should not be allowed to occur.

2. **Afforestation Policy**: A forest policy suited to the upgrading of environmental aspects should be designed. Perhaps the plants shall have to be selected not only on the consideration of quick growth but also in terms of their effects on agricultural development and environment. It is reported that some

species of plants adopted for social foresty have resulted in a considerable fall of the water-table which might threaten other forms of vegetation. Should such a thing be proved, alternative species should be developed immediately so that the harm to environment is minimised. It has often been notices that large-scale plantation takes place every year but due to poor watch and ward and practically no maintenance, sizeable part this plantation is wasted and it does no goods to the environment. Reckless exploitation of the existing forests is still going on and encroachments on forest lands are common. Strict measures shall have to be taken to check this menace. In many cases, it has been reported that due to overgrazing in the past, there exists hardly any grazing facility in the rural areas now and it is directly responsible for the failure of animal husbandry component of the I.R.D. Programmes. Forest cover is essential to the harmony of rural economy and it is to be pursued with that point of view as well.

3. **Suitable Input Mix** : Input mix that shapes a technology shall have to be designed carefully. The seeds, fertilizers, pesticides and agricultural implements have to be combined in such a way as would meet the present challenges and would also ensure stability in the future. For arriving at an optimal combination, farm education and management shall have to play an important role. Different varieties of seeds to suit different agro-climatic conditions should be developed keeping in view the experience gathered so far. Some chemical ingredients shall have to be provided to check the hardening of the upper crust of the soil and not to damage its retention capacity. As already explained, these can become big health hazards unless used properly and in desirable quantities only. Greater caution shall have to be exercised in this regard keeping in view the socio-economic conditions of the country. The use of modern agricultural implements shall have to be weighed against the prevailing employment position in the countryside. Those mechanical devices which help in deepening the agricultural operations and intensifying the labour use should developed and adopted under the TRYSEM scheme. It will be desirable to give training in modern farming as well so that more and more educated people in rural areas take to farming as a source of livelihood. Whereas the new technology should help in reducing drudgery, it should in no way result in the displacement of labour.

4. **Suitable Cropping Pattern** : A suitable cropping pattern is equally important. Environment should be allowed a free play and not made to behave every time according to the whims of man. While new cropping patterns based on market decisions will evolve themselves from time to time, by adopting suitable policy measures, the desired direction shall have to be given to them so that they help in the maintenance of environmental balance. Such exercises are warranted in social an agro-foresty as well. The plantation of those trees which

do not cause imbalance should be promoted. For example, trees affecting adversely the water level should be avoided. Researches will have to undertaken to evolve suitable cropping patterns.

5. **Rural Development Strategies** : In the ultimate analysis all rural development programmes are linked with agriculture in one or the other way. Such development activities as are likely to affect agriculture in an adverse fashion should be avoided. The activities will have to be selected according to the region specificity. If the animal husbandry activities are promoted in areas deficient in forest and grazing lands, it will lead to considerable damage to the forest and grazing lands by overuse and will upset the environmental conditions. On the other hand, rural development activities based on agricultural by-products and forest waste should be encouraged. These would be income- generating, on the one hand and help in the improvement of total environmental position of the village, on the other. Activities of health and rural sanitation should be suitably designed to encourage the environmental improvement activities. The farmers should be encouraged to raise agro-forests which will augment their incomes and simultaneously improve the environment as well.

8

The Role of India Inc. in Agricultural Development

Panduranga Venkataramulu

Corporate Social Responsibility (CSR) is an important approach through which a corporate entity tries to balances economic, social and environmental objectives in addressing stakeholder expectations and enhancing shareholder value. An essential component of corporate social responsibility is to care for the community. A positive contribution to the underprivileged communities by supporting a wide range of socio-economic, educational and health initiatives forms the core of CSR. Corporate social responsibility is a cost-effective brand building strategy with long-term orientation. It would give a good societal reputation and socially responsible identity to companies, involving the companies and their employees in the long-term process of positive social transition. The corporates throughout the world are contributing for the development of agricultural sector through their CSR initiatives. The role of India Inc. is tremendous in this regard. The present paper is an attempt to highlight some of the CSR activities of India Inc. towards the development of agriculture in India. The paper discusses the contributions made by ITC, HUL MCX and BPCL in the agricultural development.

Indian Tobacco Company (ITC)

ITC's Agri Business Division is one of India's largest exporters of agricultural commodities, has conceived e-Choupal as a more efficient supply chain aimed at delivering value to its customers around the world on a sustainable basis. The *Choupal* is a Hindi word for village square where elders meet to discuss matters of importance. ITC installs a computer with solar-charged batteries for power and a VSAT Internet connection in elected villages. The e-Choupal model has been specifically designed to tackle the challenges posed by the unique features of Indian agriculture, characterized by fragmented farms, weak infrastructure and the involvement of numerous intermediaries, among others.

The Value Chain - Farm to Factory Gate

'e-Choupal' also unshackles the potential of Indian farmer who has been trapped in a vicious cycle of low risk taking ability; low investment; low productivity; weak market orientation; low value addition; low margin and low risk taking ability. This made him and Indian agribusiness sector globally uncompetitive, despite rich and abundant natural resources. ITC's The Value Chain - Farm to Factory Gate as given in Figure 1, is a market-led business model to enhance the competitiveness of Indian agriculture and trigger a virtuous cycle of higher productivity, higher incomes, enlarged capacity for farmer risk management, larger investments and higher quality and productivity. Further, a growth in rural incomes will also unleash the latent demand for industrial goods so necessary for the continued growth of the Indian economy. This will create another virtuous cycle propelling the economy into a higher growth trajectory.

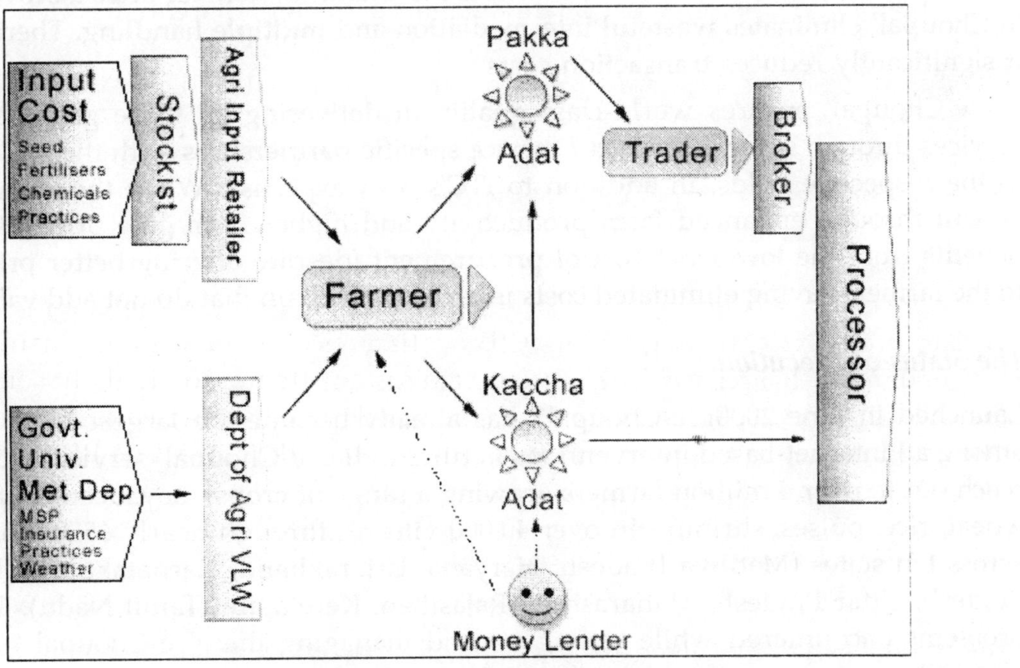

Figure 1: The Value Chain - Farm to Factory Gate

Modus Operandi of the Model: Appreciating the imperative of intermediaries in the Indian context, 'e-Choupal' leverages Information Technology to virtually cluster all the value chain participants, delivering the same benefits as vertical integration does in mature agricultural economies like the USA. 'e-Choupal' makes use of the physical transmission capabilities of current intermediaries – aggregation, logistics, counter-party risk and bridge financing – while disintermediating them from the chain of information flow and market signals.

With a judicious blend of click and mortar capabilities, village internet kiosks managed by farmers – called sanchalaks – themselves, enable the agricultural community access ready information in their local language on the weather and market prices, disseminate knowledge on scientific farm practices and risk management, facilitate the sale of farm inputs (now with embedded knowledge) and purchase farm produce from the farmers' doorsteps (decision making is now information-based).

Real-time information and customized knowledge provided by 'e-Choupal' enhances the ability of farmers to take decisions and align their farm output with market demand and secure quality and productivity. The aggregation of the demand for farm inputs from individual farmers gives them access to high quality inputs from established and reputed manufacturers at fair prices. As a direct marketing channel, virtually linked to the 'mandi' system for price discovery, 'e-Choupal' eliminates wasteful intermediation and multiple handling. Thereby it significantly reduces transaction costs.

'e-Choupal' ensures world-class quality in delivering all these goods and services through several product / service specific partnerships with the leaders in the respective fields, in addition to ITC's own expertise. While the farmers benefit through enhanced farm productivity and higher farm gate prices, ITC benefits from the lower net cost of procurement (despite offering better prices to the farmer) having eliminated costs in the supply chain that do not add value.

The Status of Execution

Launched in June 2000, 'e-Choupal', has already become the largest initiative among all Internet-based interventions in rural India. 'e-Choupal' services today reach out to over 4 million farmers growing a range of crops - soyabean, coffee, wheat, rice, pulses, shrimp - in over 40,000 villages through nearly 6500 kiosks across ten states (Madhya Pradesh, Haryana, Uttarakhand, Karnataka, Andhra Pradesh, Uttar Pradesh, Maharashtra, Rajasthan, Kerela and Tamil Nadu). The problems encountered while setting up and managing these 'e-Choupals' are primarily of infrastructural inadequacies, including power supply, telecom connectivity and bandwidth, apart from the challenge of imparting skills to the first time internet users in remote and inaccessible areas of rural India.

Several alternative and innovative solutions – some of them expensive – are being deployed to overcome these challenges e.g. Power back-up through batteries charged by Solar panels, upgrading BSNL exchanges with RNS kits, installation of VSAT equipment, Mobile Choupals, local caching of static content on website to stream in the dynamic content more efficiently, 24x7 helpdesk etc. Going forward, the roadmap includes plans to integrate bulk storage, handling and transportation facilities to improve logistics efficiencies.

ITC has taken care to involve farmers in the designing and management of the entire 'e-Choupal' initiative. The active participation of farmers in this rural initiative has created a sense of ownership in the project among the farmers. They see the 'e-Choupal' as the new age cooperative for all practical purposes.

This enthusiastic response from farmers has encouraged ITC to plan for the extension of the 'e-Choupal' initiative to altogether 15 states across India over the next few years. On the anvil are plans to channelise other services related to micro-credit, health and education through the same 'e-Choupal' infrastructure.

Another path-breaking initiative – the 'Choupal Pradarshan Khet', brings the benefits of agricultural best practices to small and marginal farmers. Backed by intensive research and knowledge, this initiative provides Agri-extension services which are qualitatively superior and involves pro-active handholding of farmers to ensure productivity gains. The services are customised to meet local conditions, ensure timely availability of farm inputs including insurance and credit, and provides a cluster of farmer schools for capturing indigenous knowledge. This initiative, which currently covers 60,000 hectares, has a multiplier impact and reaches out to 1.4 million farmers.

ITC's e-Choupal has won numerous awards some important awards are United Nations Industrial Development Organisation (UNIDO) Award at the international conference on Sharing Innovative Agribusiness Solutions 2008 at Cairo for ITC's exemplary initiatives in agri business through the e-Choupal: The Ashoka - Changemakers 'Health For All' Award 2006 for the Rural Health Services model for delivery of health services through the e-Choupals: The Stockholm Challenge 2006. This award is for using information technology for the economic development of rural communities: The inaugural 'World Business Award', instituted jointly by the International Chamber of Commerce (ICC), the HRH Prince of Wales International Business Leaders Forum (IBLF) and the United Nations Development Programme (UNDP). This award recognises companies who have made significant efforts to create sustainable livelihood opportunities and enduring wealth in developing countries.

Hindustan Unilever Limited (HUL)

Hindustan Unilever (HUL), a fast moving consumer goods company having a soap factory in a village Khamgaon, had the capital and an inclination to help farmers near its factory to be financially independent. The cost of building one check dam is over Rs. 1 lakh. Of the total rainfall here, about 55 per cent of rainfall is lost in evaporation and 35 per cent is wasted as water runs away downhill. Merely 10 per cent is ultimately used for irrigation or drinking purposes. In this scenario, a water harvesting model developed by HUL is expected to increase the water retention manifold. Having implemented the project in Parkhed, HUL hopes to more than double the area of land under cultivation in Mandka, a nearby village with an initial investment of Rs 80 lakh. It aims to get nearly 125 hectares of land owned by the farmers of Mandka under regular cultivation. It will first work on a demonstration land of 569 acres with a set of farmers to convince rest of them in Mandka of this model's utility. The model has to be built on the hilly terrain in Mandka region, which has wide variations in elevation, slope lengths and a fragile geological formation. When the water flows away down the hill, it triggers multiple problems like loss of vegetation, lack of fodder for cattle and soil erosion. The water harvesting model devised by HUL consists of building trenches and gullies, constructing check dams and wells and planting a selection of tree species that would arrest water before it flows away.

The company has taken this project to Mandka after its experiment in Parkhed. Total crop harvest has gone up crops can be grown in all seasons. The farmers, here, could grow only cotton, wheat and gram. Availability of water has helped us to grow vegetables like onions and fruits like papaya. In Parkhed, between 2003 and 2007, the company has built 1,600 trenches, 7,500 pits, 5 permanent check dams and 3 wells for a total expense of over Rs 50 lakh. According to MITTRA officials, about 2,000 people, including 120 landless farmers, have benefited from this programme and their income has gone up by three times. The cost of treating the plot was about Rs 6,000 to Rs 8,000 per hectare of land.

The company also assisted women in forming about 32 self help groups (SHGs) which have collected a total savings of Rs 3.52 lakh and helped local schools to give computer training. The migration from this village, the company says, it has reduced, with 99 per cent of the population wanting to stay back after the success of the project.

Multi Commodity Exchange (MCX)

The Multi Commodity Exchange (MCX) has tied up with India Post in Maharashtra, Madhya Pradesh and Gujarat to reach out to farmers. Under the programme, a farmer can route his query to the MCX representative stationed at the post office through the village postman. The queries are sent online to the nearest agriculture research centres and responses delivered to the farmer through the postman. Besides specific queries, support information on farm-related issues and pricing would ride on this daily network.

The exchange has set up two hubs at head post office of Jalgaon and Dhamangaon (55 km from Amravati) and one each at Itarsi (Madhya Pradesh) and Unjha (Gujarat). Another national commodity futures exchange, NCDEX, also undertakes programmes to work with farmers. ITC, a conglomerate with diversified business interest, has empowered 3.5 million farmers across 38,000 villages in nine States through its e-choupal initiative.

The MCX hubs at post offices would have a computer terminal with Internet access, a printer, a scanner, a fax machine and an electronic MCX commodity price ticker. The exchange charges Rs 10 per query on crop advisory services. "The fees do not even cover the cost of operations. Though investment on computer and accessories are one-time investments, we incur monthly expenditure in the form of rentals to the post office, computer operator salaries, etc. At many places we are using blackboard for `disseminating information," said chief economist, MCX, Mr V. Shanmugam, who heads corporate social responsibility (CSR) initiatives. Apart from making farmers understand the benefit of futures trading, we will help them get advice on fertilisers, pesticide, finance and warehousing," he said. Though the response from farmers has been slow, MCX is hopeful that it will win farmers' confidence soon. "We have received about 250 queries in six months from farmers and the numbers will definitely increase once we spread the tie-up to 1,55,000 post offices across the country," he said.

Warehousing

The MCX has also tied up 12 banks for arranging short-term finance for farmers based on warehouse receipts. Through its own subsidiary National Bulk Handling Corporation (NBHC), MCX has opened a new warehouse in Jalgaon. MCX provides daily price movements of commodities stored in the warehouses so that banks can take care of its interests. The exchange is also helping farmers get insurance cover by compiling data through farmer associations. MCX is spreading the awareness on 'timing the market' (when to sell their produce) among farmers through panchayat heads. It is also collating data of farmers through associations to help utilise products floated by the Agriculture Insurance Company.

British Petroleum Corporation Limited (BPCL)

BPCL arranges training through Pune Agricultural University, distribute seeds and fertilizers every year for agricultural upliftment. There are lectures arranged on technical guidance, use of less water for farming, teaching newer methods of agriculture, etc. Farmers are educated about fish farming and provide them all necessary information. Guidance is also given as regards buying of seeds. The methods of multiple cropping are taught. Every year farmers are given free fertilizers. Bamboo cultivation is also promoted. 5000 Bamboo saplings are planted every year, by giving it to villagers who will take care of it. Along with it, even mango, chikku and guava saplings are provided.

In the villages of Karjat, Institute of rural farmers were dependent on moneylenders for money as well as seeds and grains and thus in turn fell in the debt trap. The money lenders would harass them and exploit them. For, e.g., for every one quintal of grain borrowed, the farmer would have to return it three quintals. To avoid such exploitation, IRCED came out with the grain bank scheme. The concept of grain bank was to make grains easily accessible to the farmers and to keep them away from the clutches of the moneylenders. Grains are made available to the farmers without any advance and they also do not have to return anything extra than what they had taken. Through BPCL, there is also a small room allocated for the purpose of maintaining the grain bank in every village so that the grains do not get spoilt.

To make available water in case of scarcity, there was a need felt to start a project on rain water harvesting. The work has just begun. As this is a hilly area the problem of ground water and land water still remains. For provision of water, tube wells and small dams are made for the villagers.

Infrastructure Development

We have constructed shed for schools. For e.g. repairing the fans, replacing broken furniture, etc jetty for fishermen so that they can rest in the night when they are not fishing. We have also constructed a public toilet. We are vigilant in keeping the surroundings clean. Timely garbage disposal is also taken care of. Tree plantation programs are carried out on a regular basis.

A group of women came together in Karjat who felt the need of being together and saving some amount of money. They started meeting regularly and since then they collect savings and deposit in the bank This is a scheme whereby every village will have one woman leader. This leader is trained in such a way that she is able to manage the whole group of other women, take training sessions for them, conducts awareness programs and also provides counseling help. Such women leaders are termed as SAKHI, which means a friend in Marathi. Thus the groups serve a double purpose, that of starting a small-scale business and that of learning through various sessions. These meetings help them to improve their quality of life.

Conclusion

Corporate world needs to build a social responsibility campaign around the business model of the organisation to strengthen its brand name. Innovative practices are being implemented by corporate world to build brands in the long-run by contributing to the development of agriculture, which is not possible by the short ten second commercial advertisements. The initiatives of corporate India in this regard are vital in enabling even marginal farmers to access knowledge to compete on an equal footing in the market place. Empowerment of rural communities helps development of the economy. Further, there is much more responsibility on India Inc. to contribute for upliftment of farmers as it is benefited by the large consumer segment at the bottom of the pyramid. The blessing of farmers will definitely boost the Corporate India.

References

Multi Commodity Exchange takes Farmer's call, The Hindu Business Line, Apr 22, 2007

HUL Greens Barren Village, *Business Standard*, February 05, 2008.

http://www.itcportal.com/sets/echoupal_frameset.htm

http://www.itcportal.com/sets/rural_frameset.htm

www.bharatpetroleum.com/corporate/csr.asp

http://www.karmayog.org/csr500companies/csr500companies_8260.htm

http://www.karmayog.org/csr500companies/csr500companies_7392.htm

9

Impact of Cropping Pattern on Temporal and Spatial Changes in the Farming System of Northern Karnataka

B.P. Veerabhadrappa
S.V. Santhosh

Introduction

Indian Economy still displays overt features typical of the most underdeveloped countries of the world. It still is steered by agricultural sector, which continues to be dependent on the vicissitudes of monsoons. Low productivity is the hallmark of India's agricultural sector. Therefore the situation demands abandoning the complacence of our food front and stop claiming about the self-sufficiency in food production and efforts should be made to device the ways and means of increasing the productivity of land. Rapid agricultural development is, therefore, sine qua non for sustainable growth and development of Indian economy.

Due to continuous and intensive cultivation of land for meeting the objective of food security, the natural resources have drastically degraded over time. Many a times, this process is irreversible. Hence, such an alarming loss/degradation of natural resources must be prevented in future at all costs, otherwise the costs of recovery would be too high to imagine if left unattended. So far, studies conducted on farming systems have been few and far between. Farming systems approach in analyzing the problems of agriculture is gaining lot of importance in recent years. Such a study would throw light on the problems associated with different farming systems and enable the academicians and policy makers to formulate and implement appropriate policies for a balanced, integrated agricultural development.

This study was conducted in three dry zones of Northern Karnataka with an objective of anlayze the temporal of spatial changes in the farming systems in the selected agro-climatic zones of Northern Karnataka.

Sampling Design

For the purpose of the study three zones were selected in the northern Karnataka, namely, North Eastern Transitional Zone, North Eastern Dry Zone and Northern Dry Zone. At the rate of one district per zone, in all, three districts were selected for the study based on the highest number of taluks of a district covered by the zones.

Changes in Cropping Pattern in the Study Area

The share of area under cereals and minor millets in Zone-I declined during the all the periods with the exception of sorghum (Table-1). The decline has been quote considerable in the case of paddy, wheat and minor millets. Area under paddy declined by 32.26 per cent during period-I, 21.50 per cent during period-II and by 46.82 per cent in the overall period. The magnitude of decline in area under minor millets was still higher during period-I (-37.48%), period-II (-53.47%) and overall period (-70.91%). Similarly area under mulberry declined during all the periods. The magnitude of decline was 44.52 per cent in period-I, 62.89 per cent in period-II and 79.41 per cent in the overall period. Sorghum showed declining trend during period-I (-0.62%) to some extent, but picked up during period-II (15.48%) and overall period (14.76%). Among other cereal crops, share of area under bajra and maize increased during period-I, but declined during period-II and overall period.

The share of area under the showed increasing trend during all the periods. The magnitude of increase was 23.33 per cent, 2.78 per cent and 26.76 per cent during period-I, period-II and overall period, respectively. However, there was a sharp increase in bengalgram area (222.40%) during period-I, while there was a sharp decline during period-II (-55.59%). Interestingly, area under total oilseeds increased during all the periods. Though cotton area increased during period-I, it declined during period-II (-61.63%) and overall period (-35.55%). However, area under sugarcane increased by 25.82 per cent during period-I, while it declined by 3.02 per cent in period-II.

In the case of Zone-II, the share of area under paddy, maize, bengalgram and other pulses showed an increasing trend for all the periods, while area under crops like ragi and minor millets declined during all the periods. The proportion of area under sorghum declined during period-I, but increased during period-II (15.56%) and overall period (14.72%). Similarly area under bajra, wheat, and total cereals declined in period-I, but increased during period-II and during overall period.

The area under oilseeds increased steeply during period-I (246.89%) with a slight decline during period-II (-30.94%). Similarly, area under sugarcane increased during period-I (35.14%) but declined during period-II (-1.72%). Area under mulberry declined during all the periods.

The share of area under paddy, bajra, maize, ragi and sugarcane increased during all the periods in Zone-III. The increase in area under maize was three fold between 1997-98 and 2005-2006 (overall period). However, area under wheat, minor millets and mulberry declined. Arca under sorghum increased by 36.85 per cent during period-I, but it declined during period-II (-42.25%) and hence resulted in decline in area under the same during the over period (–25.08%). Similar trend was noticed in the case of tur and bengalgram.

Dynamics of Cropping Pattern

The Markov Chain analysis was carried out using data of ten years period (1995-96 to 2005-06) on the cropping pattern in the study area. The stability of the acreage share of the crop group aggregates considered for analysis and the transitional probability matrix capture their direction and volume of change over time. The transition probability matrix was estimated using the Minimum Absolute Deviation (MAD) estimator. The elements P_{ij} of the matrix are the conditional probabilities of the acreage share of a crop in time 't' given its share in time t-l. The diagonal elements P_{ij} (i=j), indicate the extent of stability of crops. Hence, as the diagonal elements approach zero, crop acreage become less stable, and as they approach one, the crop acreage tend to be more stable over time. The off-diagonal elements P_{ij} (i=j), are the probabilities of switching between crops. If P_{ij} is the diagonal element corresponding to the ith crop, the other elements in the ith row give the proportions of previous years acreage of ith crop it is likely to loose to other crops in the current year. The elements of the jth column gives the proportion of areas of other crops in the previous years, the ith crop is likely to gain in the current year. The results of the transitional probability matrix for crops in the selected agro climate zones of northern Karnataka are presented below.

This analysis was carried out to know the direction of cropping pattern in the Study zones. The transitional probabilities showing the extent of retention of previous under different crop groups for Zone-I, Zone-II and Zone-Ill are presented in Table 1, 2, 3.

Zone - I

The results presented in the Table-2 revealed that the highest retention of area was noticed in the case of oilseed crop group (75.83%) followed by cereals 65.67%) and pulses (56.81%). Minor millets and commercial crop group were not able to retain the magnitude of their share of previous year. Mulberry crop was able to retain its share in the cropping pattern to the extent of 1.11 per cent only.

To retain their share, cereals have gained much from the other crop groups compared to their loss to other crop groups. Pulses have gained the area to the extent of 28.68 per cent from cereals and 6.63 per cent from the pulses and at the same time it has lost its area to cereals to the extent of 40.89 per cent. Though oilseed crop group has retained its major share in the cropping pattern, it has not gained or lost much to other crop groups. Minor millets group, though not able to retain its share in the cropping pattern, it gained to the extent of 92.23 per cent from mulberry crop while, losing its area to cereals to the extent of 86.53 per cent. Commercial crop group was not able to retain its share of area, as it lost its area to cereals with little gain from other crop groups except minor millets and mulberry.

Table-1

Transitional Probability Matrix for Cropping Pattern in Zone 1

S.N	Crop Group	Total Cereals	Total Pulses	Minor Millets	Total Oilseeds	Total Commercial Crops	Mulberry
1	Total Cereals	0.6567	0.2868	0.0029	0.0081	0.0454	0.0001
2	Total Pulses	0.4089	0.5681	0.0000	0.0000	0.0230	0.0000
3	Total Minor Millets	0.8653	0.0000	0.0000	0,0000	0.0000	0.1347
4	Total Oilseeds	0.1380	0.0663	0.0000	0.7583	0.0374	0.0000
5	Commercial crops	1.0000	0.0000	0.0000	0.0000	0.0000	0.0000
6	Mulberry	0 0000	0.0000	0.9223	0.0000	0.0000	0.0777

Zone – II

Total cereals dominated the cropping pattern retaining its previous share to the extent of 62.60 per cent. The other crops that have retained their share were oilseeds (59.00%) pulses (29.90%) and mulberry (25.60%). Similarly, in this zone also, minor millets and commercial crop groups were unable to retain their share in the cropping pattern.

Table-2

Transitional Probability Matrix for Cropping Pattern in Zone II

S.N	Crop Group	Total Cereals	Total Pulses	Minor Millets	Total Oilseeds	Total Commercial Crops	Mulberry
1	Total Cereals	0.626	0.256	0.006	0.048	0.064	0.001
2	Total Pulses	0.677	0.299	0.000	0.000	0.024	0.000
3	Total Minor Millets	0.000	0.000	0.000	1.000	0.000	0.000
4	Total Oilseeds	0.049	0.361	0.000	0.590	0.000	0.000
5	Commercial crops	0.000	0.000	0.000	1.000	0.000	0.000
6	Mulberry	0 0000	0.0000	0.744	0.000	0.0000	0.256

Cereals retained their share by gaining the major area from pulses (67.70%) and to some extent from oilseeds (4.90%). Oilseeds group retained its share in the cropping pattern with higher amount of area gained from minor millet group (100%), commercial crop group (100%) and to some extent from cereals (4.80%) Minor millets group through not retained its share, gained from mulberry crop (74.40%), while commercial crop group lost its area to oilseeds with a little gain from cereals and pulses.

Zone – III

The highest retention of area was observed in cereals (71.61%) followed by oilseeds (70.50%), commercial crops (44.20%) and minor millets (25.80%). However a distinct pattern was noticed in this zone as pulses and mulberry were not able to retain their shares in the cropping pattern unlike in Zone-I and Zone-II wherein minor millets and commercial crop groups were not able to retain their share.

Table 3

Transitional Probability Matrix for Cropping Pattern in Zone III

S.N	Crop Group	Total Cereals	Total Pulses	Minor Millets	Total Oilseeds	Total Commercial Crops	Mulberry
1	Total Cereals	0.7161	0.1042	0.0109	0.0583	0.1105	0.0000
2	Total Pulses	1.0000	0.0000	0.0000	0.0000	0.0000	0.0000
3	Total Minor Millets	0.0000	0.7420	0.2580	0.0000	0.0000	0.0000
4	Total Oilseeds	0.1060	101731	0.0000	0.7050	0.0159	0.0000
5	Commercial crops	0.3685	0.1821	0.0000	0.0000	0.4420	0.0075
6	Mulberry	0 0000	0.0000	1.0000	0.0000	0.0000	0.0000

Similarly in this zone also, cereals retained their share by gaining the major area from pulses (100%) and to some extent from oilseeds (10.60%) and commercial crop groups (36.85%), without loosing much of the area to any of the other crop groups. Oilseeds group retained its share with a little gain from cereals (5.83%), whereas, commercial crop group retained its share by gaining area from cereals (11.05%) and oilseeds (1.59%). Though pulse crop group did not retain its share in the previous cropping pattern, it gained much of the area from minor millets (74.20%) and to some extent from commercial crop groups (18.21%), oilseeds (17.31%) and cereals (10.42%). Mulberry crop was not able to retain its share, but lost its area to minor millets (100%) and could not gain from any of the crop groups except commercial crops (0.75%).

Findings of the Study

The share of area under cereal crops and minor millets in Zone-I declined during all the periods with the exception of jowar. Similarly, area under mulberry declined during all the periods. To some extent jowar showed declining trend during period-I, but picked up during period-II and overall period. The share of area under tur showed increasing trend during all the periods. Interestingly, there was a sharp increase in the area under total oilseeds during all the periods. But area under cotton though increased during period-I, declined during period-II and overall period.

In the case of Zone-II, the share of area under paddy, maize, bengalgram and other pulses showed an increasing trend in all the periods, while area under ragi and minor millets declined during all the periods. Area under bajra, jowar, wheat, and total cereals declined in Period-I but increased during Period-II and during overall period.

In the case of Zone-III, share of area under paddy, bajra, maize, ragi and sugarcane increased during all the periods. However, there was a decline in wheat, minor millets and mulberry area. Area under tur and bengalgram increased during period-I, while it declined during period-II and overall period.

Suggestion

There was sharp decline in the area under one of the important crops viz., cotton due to higher incidence of pests and diseases, poor yield performance and poor quality of pesticides. Hence, integrated pest management method needs to be popularized in order to reduce cost on plant protection chemicals and supply of quality seeds by public agencies. Also there is an urgent need to protect farmers against using poor quality pesticides by strict enforcement against firms supplying such pesticides.

References

Anonymous, 2002, Economic Survey, Economic Division, Ministry of Finance, New Delhi.

Anonymous, 2003, Directory of Indian Agriculture, CMIE, New Delhi.

Buckwell, A.E., Shuchsmith, D.M. and Young, D.A., 1983, Structural projections of the Scotish dairy industry using micro and macro Marko transition matrices. Journal of Agricultural Economics, 34(1)"57-68.

Chavas, J.P. and Magand, D. 1998, A dynamic analysis of size of distribution of farms: the case of the United States dairy industry. Agribusiness, 4(4):315-329.

Deoghare, P.R., Sharma, B.M. and Goel, SIK., 1991a, Impact of mixed farming system on income and employment on small farms in Karnal district of Haryana. Agricultural Situation in India. 45(10):665-670.

Gangwar, A.C. and Ghakar, R.K., 1975, Can small scale farming yield more income? Indian Journal of Agricultural Economics, 30(3):172-181.

Johl, S.S., 1984, Some determinants of agricultural growth and development in India. Indian Journal of Agricultural Economics, 29(1):1-19.

Nanja Reddy, C., 1980, Income and employment potential of small farmers in Chennapatna block, Bangalore district, Karnataka. An unpublished Ph.D Thesis, Submitted to University of Agricultural Sciences, Banaglore.

10

Impact of Green Revolution on Agricultural Development in India

D. Pulla Rao

Introduction

Agriculture is the largest sector of economic activity in India. The significance of agriculture in India arises also from the fact that the development in agriculture is an essential condition for the development of the national economy. It provides not only food and raw materials but also employment to a very large proportion of population. It is clear, that agriculture is the backbone of the Indian economy and prosperity of agriculture can also largely stand for the prosperity of the Indian economy. Agricultural growth has direct impact on poverty eradication. Agricultural growth is also an important factor in containing inflation, raising agricultural wages and for employment generation. But after independence our agriculture was in a hopeless and deplorable condition. Our farmers were in heavy debt to the private moneylenders. They were having small and scattered holdings. They had neither the money nor the knowledge to buy and use proper equipment, good seeds and chemical fertilizers. Most of our agriculture was dependent upon rainfall and upon the vagaries of the monsoons. Productivity of land as well as of labour had been declining and was at a very low level. In spite of the fact that nearly 70 per cent of our working population was engaged in cultivation, the country was not self-sufficient in food grains. Rapid growth of agriculture is, therefore, essential to achieve the multi-pronged objectives of the Indian planning system. These include self-reliance at the national level, food security to individual households, equitable distribution of income and wealth, rapid reduction of poverty levels, absorbing the growing labour force, facilitating export growth, to achieve balance of payments equilibrium in the economy, and so on.

Agriculture and Economic Development

There are different views regarding the role of agriculture in economic development. According to some economists, if economic development begins with agriculture it cannot lead to a self sustained growth due to the following reasons:

1. There will not be sufficient capital formation.
2. It will not create the necessary climate for industrialization.
3. It cannot generate forces of expansion.

According to Nurkse, economic development consists in transferring population from agriculture to industries. But, some economists like Arthur Lewis do not completely agree with this view. According to Lewis, in France, prior to 1944, industrialization was slow because agriculture was stagnant. Agricultural development is first necessary for the following reasons:

1. Developing agriculture must first create the necessary conditions for industrialization.
2. Food grains for the growing population have to come from agriculture.
3. Raw materials for the industries have to come from agriculture.
4. Developing agriculture can create marketable surplus, which is necessary for economic development.
5. The capital- output ratio in agriculture is low and hence a small capital investment will bring large output.
6. The ratio of savings and investment can be increased if the rate of savings and investment are high in agriculture.
7. Foreign exchange is not very much necessary for agricultural development.
8. Agriculture contributes nearly 34 per cent to national income. Unless there is development in agriculture, per capita income cannot be increased, and
9. Imports are not very much necessary for agricultural development.

Therefore agricultural development should be given first priority in India. The neglect of agriculture in the second plan created several problems.

No economic reform will be successful without a steady growth in agriculture. It is estimated that if we want to achieve 5 per cent growth rate of the economy, the growth rate of agriculture should be at least 4 per cent. To achieve this growth rate in agriculture investment in the agricultural sector should be about 25 to 35 per cent of G.D.P. Thus, agricultural development is an essential precondition for economic development, especially in a predominantly agricultural country like India. If agriculture is not developed industrialization will be slow.

Objectives

The major objectives of this paper are as follows:

1. To examine the impact of green revolution on agricultural development in India.
2. To estimate the growth of agricultural production in India during the period 1960-61 to 2000-01.
3. To examine the crop wise food grain production in India during the period 1960-61 to 2000-01.
4. To estimate the increase in the production to commercial crops (cash crops) in India during the period 1960-61 to 2000-01.
5. To examine the trends of major agricultural inputs used in food grain production in India during the period 1960-61 to 1999-2000.

Collection of Data and Methodology

We have collected the required data and information from various secondary sources like Center for Monitoring Indian Economy, Handbook of Statistics on Indian Economy, and Indian Economic Survey.

Our study is limited to four decades, i.e., from 1960-61 (beginning of green revolution) to 2000-01. We have divided the entire study period in to four decades, i.e., 1st decade (1960-61 to 1970-71), 2nd decade (1970-71 to 1980-81), 3rd decade (1980-81 to 1990-91), and 4th decade (1990-91 to 2000-01). In order to estimate the growth of food grain production by area, crop wise production including commercial crops and yield and growth of major inputs used in food grain production for irrigation, fertilizer and HIV, we have calculated the per centage growth rates for each decade and for each variable and also estimated annual compound growth rates for all the above variables.

New Agricultural Strategy

The Government of India has taken several measures to increase agricultural production and productivity. These measures are responsible for the agricultural revolution called "Green Revolution". Since the mid–1960's the traditional agricultural practices are gradually being replaced by modern technology and farm practices in India. Initially, the new technology was tried in 1960-61 as a pilot project in seven districts.

The new agricultural strategy goes by the names of IADP and High Yielding Variety Programme (HYVP). The new agricultural strategy consists of:

1. Introduction of new and high yielding varieties of improved seeds.
2. Short duration crops and multiple cropping.
3. Increased use of fertilizers and pesticides.
4. Soil conservation methods.
5. Extension of irrigational facilities.
6. Adequate supply of agricultural credit and other inputs.
7. Using modern machinery and equipment.
8. Increased provision of storage, marketing and processing facilities.
9. Institutional changes like land reforms.
10. Development of infrastructural facilities.
11. Agricultural price- support.

High Yielding Variety Seeds

The Green Revolution was mainly due to hybrid seeds, which raised agricultural yields. In Mexico, Prof. Norman Borlaug developed new varieties of wheat, which were high yielding, disease-resistant and early maturing during 1960. These varieties were imported and planted in selected regions in India, which have adequate irrigation facilities. The yield from these Mexican varieties of wheat was more than 25 to 100 per cent compared to the old varieties. Encouraged by the results the government launched a programme of developing hybrid seeds suitable to Indian conditions. The Indian Council of Agricultural Research, Punjab Agricultural University, G.B. Pant Agricultural University etc., developed new seeds. Production of improved and high-yielding variety of seeds was also encouraged in Central and State government farms, etc. As a result several high-yielding variety seeds are developed. Regarding wheat we have the Mexican varieties Lerma Rojo, Sonara-64, and hybrid varieties Kalyan, P.V.18, Sharbati sona, Pusa Lerma. In rice we have TN-1, I.R.-8, Timen-3, ADT-17, Jaya, Padma etc, for maize Vijay and for bajra H.B.-1.

Introduction of such high-yielding varieties requires adequate supply of water, fertilizes, pesticides, insecticides etc. Therefore, they have to be introduced in areas where there are adequate institutional facilities under a "Package

Programme" which provides adequate supply of fertilizers, pesticides and insecticides. Areas, which are under Intensive Agricultural Area Programme (IAAP) and other areas, which are suitable are selected for introducing High yielding Varieties Programme (HYVP). In addition to the High Yielding Variety Programme (HYVP) the Government has taken several steps to increase agricultural production and remove the bottlenecks in agricultural progress. It has adopted a new agricultural strategy. The other aspects are short duration crops, multiple cropping, machinery, extension of irrigation, price support, processing, storage and marketing facilities, improved credit facilities, institutional changes, production of supplementary food products etc.

Thus, the new Agricultural strategy, which has led to revolutionary changes in agriculture, has broken agricultural stagnation in India and resulted in higher productivity that lead to higher production. Green Revolution is a technological break through in Indian agriculture.

Progress Achieved

Green revolution resulted in a record gain of output from 51 million tonnes in 1950-51 to 131 million tonnes in 1978-79 and thus made India as one of the biggest agricultural producers in the world. India also became an exporter of food grains around that time. Yield per unit of Net Sown Area (NSA) improved by more than 30 per cent between 1947 and 1979, when the green revolution was considered to have delivered its goods fully. The crop area under high yielding varieties grew from 7 per cent to 22 per cent of the total cultivated area during the 10 years of the green revolution. During this period, the area of HYV was covered by wheat (around 70%), rice, millet and some other crops. Crop area under HYV needed more water, fertilizers, pesticides, fungicides and certain other chemicals. This spurred the growth of the local manufacturing sector and helped the Indian economy in terms of employment generation and its contribution to the country's GDP.

Trends of Food grain Production in Agriculture

Food grain production occupies the most dominant position in India's agriculture, which covers over 65 per cent of the Gross Cropped Area (GCA). Since the beginning of the green revolution in the mid 1960s, the country has shown quite impressive growth in food grain production. Chronic food deficits and hovering Malthusian crisis were overcome and that have given way to self-sufficiency in food grains and occasional marginal surpluses by the 1980s and 1990s. However, with the population growing at nearly 2 per cent per year (nearly 18 to 20 million people being added every year) and accelerating income growth, demand for food is continuously growing in the economy. India's food grain production has also been increasing rapidly over the years.

It is estimated, that India's food grain production in the early phase of green revolution has decreased from 82,018 million tonnes in 1960-61 to 72.35 million tonnes in 1965-66 and then increased to 108.42 million tonnes in 1970-71. The food grains production further increased to 121.03 million tonnes in 1975-76 and to 150.44 million tonnes in 1985-86. However, during the early phase of economic reforms (1990-91), the food grain production was 176.39 million tonnes and ultimately it increased to 199.54 million tonnes in 2000-01. Thus, the amount of food grain production during the period 1960-2001 has grown more than two-fold. The per centage growth rate of the food grain production during first decade was 32.19; 19.52 in the second decade; 36.11 in the third decade; and 13.12 in the last decade. The per centage growth rate of food grain production was the highest in the third decade. The growth rate of food grain production was the lowest in the fourth decade. This was probably due to the erratic nature of monsoon in the country. The annual compound growth rate of the food grain production was 2.25 per cent. This suggests that economic reforms had a positive impact on green revolution for increasing the food grains production in the country. Yield of food grains per hectare of net sown area also increased substantially during our study period (except in 1965-66). Yield of food grains in 1960-61 was 710 kg/ha and decreased to 629 kg/ha in 1965-66. The yield of food grains increased to 872 kg/ha in 1970-71 and continuously increased and ultimately it reached to 1648 kg/ha in 2000-01. Yield of food grains per hectare during the period 1960-2001 has grown more than two-fold. The per centage growth rate of the yield of food grains in the first decade of the study period (1960-61 to 1970-71) was 22.82 and second decade it was 17.32 and in third decade it was 34.90 and in the fourth decade it was 19.42. The per centage growth rate of the yield of food grains reached the highest in the third decade of our study period (i.e., 1980-81 to 1990-91). The annual compound growth rate of the yield of food grains was 2.13 per cent. There has also been a phenomenal rise in the size of the cultivable area. In 1960-61 the total amount of cultivable land was 115.58 million hectares, and it slightly decreased to 115.10 million hectares in 1965-66 and increased to 124.32 million hectares in 1970-71 and increased to 128.02 million hectares in 1985-86 and continuously decreased and ultimately stood at 121.05 million hectares in 2000-01. This was probably due to the erratic nature of monsoon and lack of proper irrigational facilities. The per centage growth rate of cultivable area in the first decade was 7.56 and second decade it was 1.89 and third decade it was only 0.92. In the fourth decade it dropped to an all time low of -5.31. This was probably due to the erratic nature of monsoon and lack of proper irrigational facilities in the county. The annual compound growth rate of the cultivable land was only 0.12 per cent. This growth rate is much less than the annual growth rates of production and yield of food grains in the study period (Please see Table-1).

Food Grains Production by Crop

The major achievement of the new strategy is to boost the production of major cereals like rice and wheat. Table-2 gives the production of the principal food crops during the last four decades. A close look at the table reveals the increase in rice production from 34.58 million tonnes in 1960-61 to 87.70 million tonnes in 2000-01, signifying a break through in this major crop of India. The per centage growth rate of the production of rice in the first decade was 22.09; in the second decade it was 27.02; in the third decade it was 38.52; and it was 18.05 in

Table -1

Growth of Food grain Production in India: 1960-61 to 2000-01

Year	Area (Million hectares)	Production (Million tonnes)	Yield (kg/hectare)
1960-61	155.58	82.02	710
1965-66	115.10	72.35	629
1970-71	124.32	108.42	872
1975-76	128.18	121.03	944
1980-81	126.67	129.59	1023
1985-86	128.02	150.44	1175
1990-91	127.84	176.39	1380
1995-96	121.01	180.42	1491
2000-01	121.05	199.54	1648
Growth Rates			
1st decade (1960-61to1970-71)	7.56	32.19	22.82
2nd decade (1970-71to1980-81)	1.89	19.52	17.32
3rd decade (1980-81to1990-91)	0.92	36.11	34.90
4th decade (1990-91to2000-01)	-5.31	13.12	19.42
Annual compound Growth rate	0.12	2.25	2.13

Source: Indian Economic Survey.

the last decade. The annual compound growth rate of the production of rice in the study period was 2.35 per cent.

The production of wheat which stood at 11 million tonnes in 1960-61 rose to 36.31 million tonnes in 1980-81 and further rose to 69.68 million tonnes in 2000-01. The per centage growth rate of the production of wheat in the first decade was 116.64; in the second decade it was 52.37; in the third decade it was 51.86; and it was 26.37 in the fourth decade. The annual compound growth rate of the production of wheat was 4.72 per cent.

The production of coarse cereals was 23.74 million tonnes in 1960-61 rose to 29.02 million tonnes in 1980-81 and further increased to 31.08 million tonnes in 2000-01. The per centage growth rate of production of coarse cereals was 28.68 in the first decade; it was -5.00 in the second decade; 12.68 in the third decade; and it was -4.95 in the last decade. The annual compound growth rate of production of the coarse cereals was only 0.67 per cent.

The production of pulses in our country is in an alarmic condition. The production of pulses was 12.70 million tonnes in 1960-61 rose to 14.26 million tonnes in 1990-91 then gradually decreased to 11.08 million tonnes in 2000-01. The per centage growth rate of pulses in the first decade of our study period was -6.93; it was -10.07 in the second decade; 34.15 in the third decade; and -22.30 in the fourth decade. Unfortunately the annual compound growth rate of pulses recorded a negative growth (-0.34 per cent).

The production of total food grains was 82.02 million tonnes in 1960-61 rose to 129.59 million tonnes in 1980-81 and then gradually increased to 199.54 million tonnes in 2000-01.The per centage growth rate of total food grains in the first decade of the study period was 32.19; it was 19.52 in the second decade; 36.11 in the third decade; and13.12 in the fourth decade. The annual compound growth rate of total food grains production was 2.25 per cent (Please see Table-2).

Growth in the Production of Commercial Crops

Green revolution was mainly directed to growth in the production of food grains. It did not affect initially the production of commercial crops or cash crops such as sugarcane, cotton, jute, oilseeds and tobacco; these crops did not record any significant improvement between 1960-61 and 1973-74. The production of sugarcane was recorded 110 million tonnes in 1960-61 and increased to 126.37 million tonnes in 1970-71 and then increased to 295.96 million tonnes in 2000-01. However, significant improvement in the output of sugarcane, took place after 1973-74. The per centage growth rate of sugarcane production in the first decade was 14.88; 22.06 in the second decade; 56.27 in the third decade; and

Table-2
Food grain Production by Crop in India: 1960-61 to 2000-01

(Million tonnes)

Year	Cereals			Pulses	Total Food Grains
	Rice	Wheat	Coarse Cereals		
1960-61	34.58	11.00	23.74	12.70	82.02
1965-66	30.59	10.40	21.42	9.94	72.35
1970-71	42.22	23.83	30.55	11.82	108.42
1975-76	48.74	28.84	30.41	13.04	121.03
1980-81	53.63	36.31	29.02	10.63	129.59
1985-86	63.83	47.05	26.20	13.36	150.44
1990-91	74.29	55.14	32.70	14.26	176.39
1995-96	76.98	62.10	29.03	12.31	180.42
2000-01	87.70	69.68	31.08	11.08	199.54
Growth Rates					
1st decade (1960-61 to 1970-71)	22.09	116.64	28.68	-6.93	32.19
2nd decade (1970-71 to 1980-81)	27.02	52.37	-5.00	-10.07	19.52
3rd decade (1980-81 to 1990-91)	38.52	51.86	12.68	34.15	36.11
4th decade (1990-91 to 2000-01)	18.05	26.37	-4.95	-22.30	13.12
Annual compound Growth rate	2.35	4.72	0.67	-0.34	2.25

Source: Indian Economic Survey.

22.78 in the last decade of the study period. The annual compound growth rate of sugarcane production was 2.50 per cent.

In the case of production of cotton it was 5.60 million bales in 1960-61; 4.85 million bales in 1965-66; 12.86 million bales in 1995-96; and 9.52 million bales in 2000-01. The per centage growth rate in the production of cotton was recorded -15.00 in the first decade; 47.27 in the second decade; 40.37 in the third decade; and -3.25 in the last decade of the study period. The annual compound growth rate of cotton production was 1.33 per cent.

In the production of Jute it was 5.26 million bales in 1960-61. It has gone up to 12.65 million bales in 1985-86; but decreased to 10.56 million bales in 2000-01. The per centage growth rate of production of Jute was recorded 17.68 in first decade; 31.82 in second decade; 13.11 in the third decade; and 14.41 in the last decade of the study period. The annual compound growth rate of the Jute production was 1.76 per cent.

There was considerable improvement in the production of oilseeds but the improvement was not such as to be called a revolution. The production of oilseeds was 6.98 million tonnes in 1960-61 and increased to 10.83 million tonnes in 1985-86 and further increased and stood at 18.44 million tonnes in 2000-01. The per centage growth rate of oilseeds production was recorded 37.96 in first decade; -2.70 in the second decade; 98.61 in third decade; and -0.91 in the last decade of the study period. The annual compound growth rate of oilseeds production was 2.46 per cent.

Tobacco is one of the major commercial crops in our country. The production of tobacco was not increased substantially during the period under study. The production of tobacco was 0.31 million tonnes in 1960-61; 0.48 million tonnes in 1980-81; and 0.49 million tonnes in 2000-01. The per centage growth rate of tobacco production was 16.13 in the first decade; 33.33 in the second decade; 16.67 in the third decade; and -12.50 in the last decade of the study. The annual compound growth rate of tobacco production was 1.15 per cent (Please see Table 3).

Growth of Inputs Used in Indian Agriculture

Fertilizer, HYV, irrigation and rainfall are treated as the major components for determining the production/productivity in Indian agriculture. The trends of the above inputs have increased over the years except rainfall, which exercises large-scale regional-variation with respect to time span.

It is to be noted here that during the starting period of Green Revolution (1960-61), the amount of irrigation [measured as Net Irrigated Area (NIA) as a per centage of Net Sown Area (NSA)] and fertilizer [(measured as per hectare consumption of chemical fertilizer (NPK)] were just 18.51 per cent and 1.91 kg. However, both increased to 19.44 per cent and 9.4 kg in 1967-68; and 32.81 per cent and 63.5 kg in 1989-90. During the early phase of economic reforms (1990-91), the figures were 33.58 per cent and 67.5 kg and then increased to 39.21 per cent and 88.85 kg in 1999-2000. The per centage growth rate of the irrigation in 1999-2000 over 1960-61 was 111.83. The annual compound growth rate of the irrigation was 1.89 per cent. The per centage growth rate of the fertilizer consumption in 1999-2000 over 1960-61 was tremendously increased to 4,551.83.

Table – 3

Production of Commercial Crops in India: 1960-61to 2000-01

(Million tonnes)

Year	Crops				
	Sugarcane	**Cotton (1)**	**Jute (2)**	**Oilseeds (3)**	**Tobacco**
1960-61	110.00	5.60	5.26	6.98	0.31
1965-66	123.99	4.85	5.78	6.40	0.29
1970-71	126.37	4.76	6.19	9.63	0.36
1975-76	140.60	5.95	5.91	10.61	0.35
1980-81	154.25	7.01	8.16	9.37	0.48
1985-86	170.65	8.73	12.65	10.83	0.44
1990-91	241.05	9.84	9.23	18.61	0.56
1995-96	281.10	12.86	8.81	22.11	0.54
2000-01	295.96	9.52	10.56	18.44	0.49
Growth Rates					
1st decade (1960-61 to 1970-71)	14.88	-15.00	17.68	37.96	16.13
2nd decade (1970-71 to 1980-81)	22.06	47.27	31.82	-2.70	33.33
3rd decade (1980-81 to 1990-91)	56.27	40.37	13.11	98.61	16.67
4th decade (1990-91 to 2000-01)	22.78	-3.25	14.41	-0.91	-12.50
Annual compound Growth rate	2.50	1.33	1.76	2.46	1.15

Note: (1): Production in million bales of 170 kg. each.
(2): Production in million bales of 180 kg. each.
(3): For nine Oilseeds out of eleven in all.

Source: Indian Economic Survey.

The annual compound growth rate of the fertilizer consumption was 10.07 per cent.

Corresponding to the figures of irrigation and fertilizer, the coverage of HYV (measured as a per centage of area under food crops) and rainfall (measured as actual rainfall as a per centage of normal rainfall) were 41 per cent and 113 per cent in1983-84. They increased to 48.3 per cent and 101 per cent in 1989-90. However, during the early phase of economic reforms, these figures were 50.9 per cent and 106 per cent, which increased to 63.5 per cent (HYV) and decreased to 96 per cent (rainfall) in 1999-2000. The coverage of HYV in 1999-2000 over 1983-84 was 54.88. The annual compound growth rate of the coverage of HYV was 2.77 per cent (Please see Table 4).

Conclusions and Suggestions

Agriculture is a major sector in India and continuous to be the lifeline for the Indian economy. Rapid growth of agriculture is, therefore essential to achieve the multi-pronged objectives of the Indian planning system. The government of India has taken several measures to increase agricultural production and productivity. These measures are responsible for the agricultural revolution, which is called "Green Revolution". The green revolution was mainly due to hybrid seeds, which raised agricultural yields. In addition to the High Yielding Variety Programme (HYVP) the government has taken several steps to increase agricultural production. The other aspects are short duration crops, multiple cropping, farm mechanization, extension of irrigation, price support, processing, storage and marketing facilities, improved credit facilities, institutional changes, production

Table-4
Growth of Major inputs used in Food grain production in India: 1960-61 to 1999-2000

Year	Irrigation (a)	Fertilizer (b)	HYV (c)	Rainfall (d)
1960-61	18.51	1.91	NA	-
1967-68	19.44	9.40	NA	-
1983-84	29.37	42.9	41.0	113
1989-90	32.81	63.5	48.3	101
1990-91	33.58	67.5	50.9	106
1991-92	35.21	69.8	53.1	91
1999-2000	39.21	88.85	63.5	96
% Growth rate in 1999-2000 over four decades	111.83	4,551.83	54.88	
Annual compound growth rate	1.89	10.07	2.77	

Note: (a) Irrigation: NIA as a % of NSA; (b) Fertilizer: Per hectare consumption of chemical fertilizer Kg/ha; (c) HYV: HYV area as % of area under food crops, and (d) Rainfall: Actual rainfall as percentage of normal rainfall.

Source: Indian Economic Survey.

of supplementary food products etc. The new agricultural strategy has led to revolutionary changes in agriculture.

India's food grain production has increased from 82.02 million tonnes in 1960-61 to 199.54 million tonnes in 2000-01. Yield of food grains per hectare also increased substantially from 710 Kg/ha in 1960-61 to 1648 Kg /ha in 2000-01. But the area of cultivable land was increased from 155.58 million hectares in 1960-61 to only 121.05 million hectares in 2000-01. The area under cultivable land has not much increased as compared to production and yield during the study period. There in a much increase in the cultivable land in the first decade of the study period. But the production and yield also increased substantially in third decade of the study.

Production of rice was increased substantially in third decade, wheat and coarse cereals in first decade and pulses in the 3rd decade. Ultimately the total food grains increased much higher in the third decade than in the other decades of our study period.

In the case of commercial crops, sugarcane and oil seeds were increased much higher in the third decade; cotton, tobacco and jute were increased substantially in the second decade. The above analysis clearly suggests that the adoption of chemical fertilizer is the main factor, which determines the productivity of agriculture in the Indian economy. However, it has to be supported by other important agricultural inputs like irrigation, HYV, rainfall etc. In the face of positive impact of fertilizer on agricultural productivity, we are indeed faced with a more challenging task. This is because with a view to increase food grain production and other planning objectives, we might lean upon the use of more fertilizers and more pesticides, which would have adverse effects on ecology, environment and health conditions of both human beings and animal species on earth. Moreover, the present use of resources should not deplete the resource-base to such an extent that the development process becomes dependent and is not sustainable in the future. There are some suggestions to improve the Indian agriculture without undue dependence on chemical fertilizer:

1. New agricultural strategies are almost confined to food grains. It must be extended to non food grains as well. It must be extended to other agricultural crops also.
2. Irrigation facilities must be increased.
3. There must be revolutionary changes in the methods of cultivation of small farmers.
4. Extension of new agricultural strategies to dry farming areas must be taken up.

Nearly 70 per cent of cultivated land in India is still under dry farming because they do not have irrigational facilities. Construction of minor-irrigation projects, scientific and optimum use of available water resource in these areas, development of new methods of water utilization etc. are some of the steps to extend Green Revolution to dry areas. In other worlds, effective programme has to be launched for "green new areas".

References

Bhalla, G.S. and Singh Gurmail. (2001): Indian Agriculture-Four decades of Development, Sage Publications, New Delhi.

Government of India. (2001):"National Agricultural Policy", Yojana, Vol. 45, No.1, 4-11.

Handbook of Statistics on Indian Economy. (2001): Reserve Bank of India, Ministry of Agriculture, Government of India, New Delhi.

Iqbal, C. (1979): "High Yielding Varieties of Seeds and their Impact on Agricultural Development", in A. Mohammad (ed.) book Dynamics of Agricultural Development in India, Concept Publications, New Delhi, India, 71-80.

Mohammad, A. (1979): Dynamics of Agricultural Development in India, Concept Publications, New Delhi, India.

Rao. C.H.H. (2001): "Technological Change in Indian Agriculture: Emerging Trends and Perspectives", in K.S. Dhindsa and A. Sharma, (ed.), Dynamics of Agricultural Development, Concept Publications, New Delhi, Vol. 2, 31-49.

Singh, K.K., Khan, M. and Shekhawat, M.S. (2000), "Green Revolution - How Green it is?", Yojana, Vol. 44, No. 6, 26-28.

Tilak, J.B.G. (1993): "Education and Agricultural Productivity in Asia: A Review", Indian Journal of Agricultural Economics, Vol. 48, No. 2, 187-200.

11

Sustainable Agriculture Development: Effects & Strategies

S.K. Dhage & B.G. Lobo

Introduction

The term sustainable agriculture means an integrated system of plant and animal production practices, satisfy human food and fiber needs, enhance environmental quality and the natural resource base on which the agricultural economy depends. It makes the most efficient use of nonrenewable resources and on-farm resources and integrates, where appropriate, natural biological cycles enhance the quality of life for farmers and society as a whole over the long term. Over the past three decades, the concerns about conventional, technology-based, and energy-intensive agriculture are deepening, while the interest in sustainable agriculture is increasing rapidly.

Ecologically and environmentally, the continued use of chemical fertilizers has increased soil erosion and decline soil productivity. Increased resistance of weeds and insects to herbicides and insecticides, combined with the destruction of wildlife and beneficial insects by pesticides may trigger a new vicious cycle both economically and ecologically.

The intensive use of chemicals in agriculture soil and water are more and more contaminated. From the natural resources perspective, fossil fuel and other chemicals are non-renewable, thus, the depletion of finite reserves of concentrated plants nutrients is only a matter of time. At present and in foreseeable future, there are no substitutes that can replace these depleted resources. Therefore, it seems sensible to consider alternative approaches, like sustainable agriculture. Following are the negative effects of the conventional agriculture which is the root of the problem

Negative effects of Conventional Agricultural Practices

Conventional farming practices has the characteristics as rapid technological innovation; large capital investments in order to apply production and management technology; large-scale farms; single crops/row crops grown continuously over many seasons; uniform high-yield hybrid crops; extensive use of pesticides, fertilizers, and external energy inputs; high labor efficiency; and dependency on agribusiness. Because of which this system has following negative effects:

1. **Decline in soil productivity**: Decline in soil productivity can be due to wind and water erosion of exposed topsoil; soil compaction; loss of soil organic matter, water holding capacity, and biological activity; and sanitation of soils and irrigation water in irrigated farming areas.
2. **Use of Chemical Fertilizers**: Ecologically and environmentally, the continued use of chemical fertilizers has increased soil erosion and decline soil productivity. Increased resistance of weeds and insects to herbicides and insecticides, combined with the destruction of wildlife and beneficial insects by pesticides may trigger a new vicious cycle both economically and ecologically. As a result of the intensive use of chemicals in agriculture soil and water are more and more contaminated, human beings are faced with the dilemma of consuming unsafe food and drinking polluted water, or spending extra money to restore the damaged environment.
3. **Water Pollutants and Water Scarcity**: Agriculture is the largest single non-point source of water pollutants including sediments, salts, fertilizers (nitrates and phosphorus), pesticides, and manures. Pesticides from every chemical class have been detected in groundwater and are commonly found in groundwater beneath agricultural areas; they are widespread in the nation's surface waters. It creates 'dead zones' due to nutrient runoff affect many rivers, lakes, and oceans. Reduced water quality impacts agricultural production, drinking water supplies, and fishery production. Water scarcity in many places is due to overuse of surface and ground water for irrigation with little concern for the natural cycle that maintains stable water availability.
4. **Global Climate Change**: Agriculture's link to global climate change is just beginning to be appreciated. Destruction of tropical forests and other native vegetation for agricultural production has a role in elevated levels

of carbon dioxide and other greenhouse gases. At the same time, given its vulnerability to climate and weather, agriculture's main challenge is adapting to the effects of climate change. Higher temperatures and an intensification of the hydrological cycle will bring both more water evaporation and more precipitation, which will be very unevenly distributed. Rainfall and cropping patterns will be altered and areas exposed to vector-borne plant and animal diseases will expand. The vulnerability of agriculture to severe meteorological events will increase

5. **Health Hazards**: Potential health hazards are tied to sub-therapeutic use of antibiotics in animal production, and pesticide and nitrate contamination of water and food. Farm workers are poisoned in fields, toxic residues are found on foods, and certain human and animal diseases have developed resistance to currently used antibiotics.
6. **Desertification**: A major cause of desertification has been and remains inappropriate land use, especially cultivation of fragile and marginal soils, overgrazing of pasture, inadequate soil conservation practices and deforestation. Soil degradation reduces agricultural productivity and ultimately renders land unusable for agriculture. There are also negative spillovers: for example, flows of off-farm sediment increase the cost of providing clean drinking water, because silting in rivers and reservoirs, damage roads and buildings and degrade aquatic ecosystems.
7. **Other elements**: Other environmental ills that have become resistant to one or more pesticides; stresses on pollinator and other beneficial species through pesticide use; loss of wetlands and wildlife habitat; and reduced genetic diversity due to reliance on genetic uniformity in most crops and livestock breeds.

Strategies of Sustainable Agricultural Development

Today's economic growth is impoverishing growth or unsustainable growth is a type of economic growth when the economy has grown in quantitative terms but the economy's reproductive capacity has declined because of environmental and natural resource degradation and other associated problems in the economy. Following strategies can be suggested to the long term sustainable development:

1. **Crop Rotations & Pest Management Techniques** : It mitigate weeds, disease, insect and other pest problems; provide alternative sources of soil nitrogen; reduce soil erosion; and reduce risk of water contamination by agricultural chemicals pest control strategies that are not harmful to natural systems, farmers, their neighbors, or consumers. Which reduce the need

for pesticides by practices such as scouting, use of resistant cultivars, timing of planting, and biological pest controls increased mechanical/ biological weed control; more soil and water conservation practices; and strategic use of animal and green manures use of natural or synthetic inputs in a way that poses no significant hazard to man, animals, or the environment.

2. **Group based Technologies**: More and more emphasis will need to be placed on the group based technologies. The areas where group action is called for include:

   (a) Synchronization of sowing schedules of certain crop to avoid synchronization with pest reproduction cycle in certain crops where this is a major problem.

   (b) Watershed management of both arable and non-arable land belonging to individuals as well as villages, forest and revenue department.

   (c) Synchronized sprays compel with pest reducing crop rotations and mixtures to minimize use of chemicals in the short run and total elimination in the long run.

   (d) Biological pest control coupled with water management and drainage.

3. **Agro-forestry and Rainwater Harvesting**: The shift from crop to trees (horticultural and/or timber) is taking place both for reducing the need for outside labour and also for reaping larger commercial gain by larger farmers. Agro-forestry for high and slow growth regions thus is an urgent priority. In many parts of the world, there is renewed interest in the traditional practice of rainwater harvesting as a method of combating increasing water scarcity. Water harvesting can be effective in arid, semi-arid and semi humid areas where surface or ground water supplies are not available or are uneconomical to develop. Run-off water can be collected over large areas (macro-catchments), from micro-catchments or in the form of floodwater, and is then stored in reservoirs, cisterns or in the soil. Adoption of this technology can allow crop production in areas where otherwise it would not be possible, reduce the risk of crop failure and generally increase yields in rain fed agriculture.

4. **Development of drought prone regions & Land Development** : The continued neglect of dry regions is really criminal. Hill areas or drought prone regions which are depend on domestic servant or a cheap labour. National Commission on Development of Backward Areas (1981, Planning Commission, New Delhi) went so far as to say that we should not try to create conditions by which supply of cheap labour for large irrigation

projects is affected adversely. Need to maintain soil fertility and water purity, conservation and improvement the chemical, physical and biological qualities of the soil, recycling of natural resources and conserving energy. Sustainable agriculture produces diverse forms of high quality foods, fibers and medicines.

5. **Integration of Human, Science & Environment**: Need to use locally available renewable resources, appropriate and affordable technologies and minimizes the use of external and purchased inputs, thereby increasing local independence and self sufficiency and insuring a source of stable income for peasants, family and small farmers and rural communities. This allows more people to stay on the land, strengthens rural communities and integrates humans with their environment. Sustainable agriculture respects the ecological principles of diversity and interdependence and uses the insights of modern science to improve rather than displace the traditional wisdom accumulated over centuries by innumerable farmers around the world.
6. **Equitable and Participatory vision of development**: Sustainable agriculture is a model of social and economic organization based on an equitable and participatory vision of development which recognizes the environment and natural resources as the foundation of economic activity. Agriculture is sustainable when it is ecologically sound, economically viable, socially just, culturally appropriate and based on a holistic scientific approach.
7. **Forest Management**: The key point of forest management is to convert resource potential into economical values, without affecting sustainability. Present problems are: too much high-quality timber is used as fuel, and brushes in marginal areas are collected exhaustively (by pulling roots, rather than cutting stems). Therefore, suggestions are a) Save good timber from fuel wood collecting and alternative sources of energy must be found first. b) Save brushes in marginal land and planted fuel forest is strongly recommended to improve the way of collecting firewood. c) Policy and regulation of forest laws to implement and enforce the law. The first priority is to set a forest price, and keep all licensed logging under control.
8. **Climate change activities**: For facing the challenge of climate change following steps can be adopted:

   (a) making climate information more relevant and usable;

   (b) developing appropriate tools for prioritising responses;

(c) applying climate risk screening tools at the project level;

(d) identifying and using appropriate entry points for climate information in project planning;

(e) shifting emphasis to implementation, as opposed to developing new plans; and

(f) encouraging meaningful co-ordination and the sharing of good practices.

The strategies suggested above require strong social and economic capabilities to be implemented effectively.

Conditions for Sustainable Development

Agricultural can only be sustainable if the conditions are fulfilled shown in the Fig. 1. Supportive policies by government, participation of individuals & local groups, external institutions & NGOs, environment friendly technology, global Responsibility, community based programme and build local access of inputs are the condition on which success can be define as to manage & maintain, conservation, rehabilitation of the sustainable development.

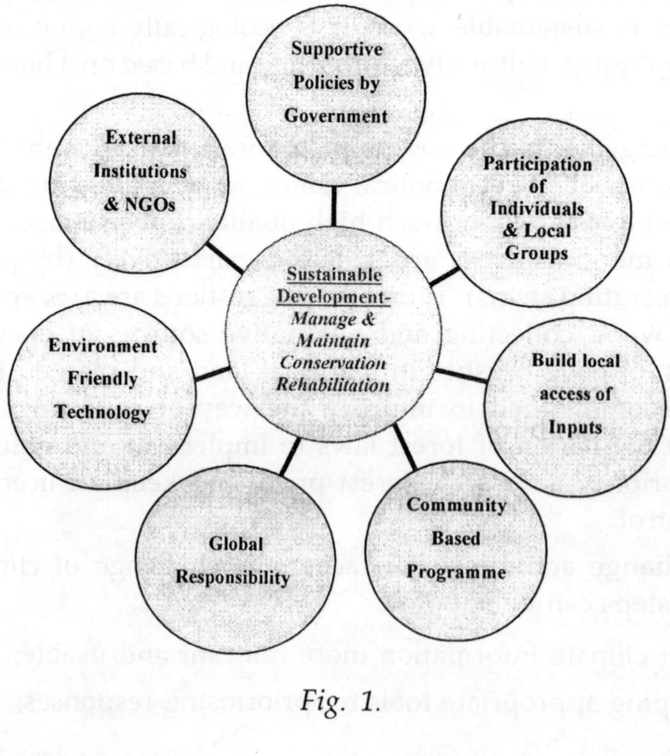

Fig. 1.

Conclusion

Many in the agricultural community have adopted the sense of urgency and direction pointed to by the sustainable agriculture concept. Lack of sharp definition has not lessened its authenticity. Sustainability has become an integral component of many government, commercial, and non-profit agriculture research efforts, and it is beginning to be woven into agricultural policy. Increasing numbers of farmers and ranchers have embarked on their own paths to sustainability, incorporating integrated and innovative approaches into their own enterprises.

As Youngberg and Harwood's 1989 statement still holds true: "We are yet a long way from knowing just what methods and systems in diverse locations will really lead to sustainability. In many regions of the country, however, and for many crops, the particular mix of methods that will allow curtailing use of harmful farm chemicals or building crop diversity, while also providing economic success, are not yet clear. The stage is set for challenging not only farm practitioners, but also researchers, educators, and farm industry." Development of a productive but sustainable agricultural system would require initiatives on many fronts. Transgenic technologies can contribute in a limited but significant way to the lofty goal of sustainable agriculture.

Reference

live A. Edwards, Rattan Lal, Patrick Madden, Robert H. Miller, and Gar House, editors, Sustainable Agricultural Systems (Soil and Water Conservation Society, 1989). Chapters 1-6 and Chapter 38.

Food, Agriculture, Conservation, and Trade Act of 1990 (FACTA), Public Law 101-624, Title XVI, Subtitle A, Section 1603 (Government Printing Office, Washington, DC, 1990) NAL Call # KF1692.A31 1990.

Garth Youngberg and Richard Harwood, "Sustainable Farming Systems: Needs and Opportunities," American Journal of Alternative Agriculture (1989) 4(3 & 4): p. 100.

International Food Policy Research Institute (IFPRI). 1995. A 2020 vision for food, agriculture, and the environment: The vision, challenge, and recommended action. Washington, D.C.: IFPRI.

J. Patrick Madden and Scott G. Chaplowe, editors, For All Generations: Making World Agriculture More Sustainable (World Sustainable Agriculture Association, 1996). Chapter 1 and Chapter 10.

NGO Sustainable Agriculture Treaty (Global Forum at Rio de Janeiro, June 1-15, 1992). Available at Information Habitat Website: http://habitat.igc.org/treaties/at-20.htm (8/23/07).

Stephen R. Gliessman, "An Ecological Definition of Sustainable Agriculture," Principles of Agro ecology and Sustainability *(1998)*. Available at Agro ecology Home Website: http://agroecology.org/Principles_Def.html (6/9/08)

Swaminathan, M. S., Sustainable Agriculture: Towards an Evergreen Revolution, Konark Publishers, Delhi, 1996.

World Commission on Environment and Development (1987) Our Common Future, Oxford University Press, Oxford.

12

Sustainable Agricultural Development: Problems and Policies

B.G. Lobo and S.K. Dhage

Introduction

Sustainable development has given rise to various visions of the world of the future, of possible tradeoffs and of externalities. Through volunteerism and involvement, people are becoming more active in creating healthy, sustainable communities at the neighborhood, municipal and regional levels within both geographical communities and communities of interest. This greater activity in communities and in the social fabric leads to an interest in examining the social impact of developments. It is defined as agriculture that balances the need for essential agricultural commodities such as food, fibre, etc. with the necessity of protecting the physical environment and public health, the foundation of agriculture. The definition is not intended to be highly specific, but to imply a multi-dimensional optimization of potentially competing forces, with a need for a holistic look at agricultural production at local, national as well as global levels and at all the various tradeoffs involved in promoting sustainable development, that will nonetheless assure an adequate supply of agricultural products for the world's population for as long as possible.

Sustainable agricultural development is one of the prime objectives in all countries in the world, whether developed or developing. The broad objective of sustainable agriculture is to balance the inherent land resource with crop requirements, paying special attention to optimization of resource use towards achievement of sustained productivity over a long period. Sustainable agricultural

sustainable development increase in crop production could be achieved by adopting a variety of agricultural technologies by improving crop management technology through the use of high yielding, input responsive and soil, climatic and biotic stresses - tolerant crop varieties.

Before going to the detail analysis of problems and policies I would like to explain comparison of characteristics of 'Sustainable Agriculture' and 'Conventional Agriculture'

Problems of Sustainable Agricultural Development

Ecologically and environmentally, the continued use of chemical fertilizers has increased soil erosion and decline soil productivity. Increased resistance of weeds and insects to herbicides and insecticides, combined with the destruction of wildlife and beneficial insects by pesticides may trigger a new vicious cycle both economically and ecologically. As a result of the intensive use of chemicals in agriculture soil and water are more and more contaminated, human beings are faced with the dilemma of consuming unsafe food and drinking polluted water, or spending extra money to restore the damaged environment. From the natural resources perspective, fossil fuel and other chemicals are non-renewable, thus, the depletion of finite reserves of concentrated plants nutrients is only a matter of time. At present and in foreseeable future, there are no substitutes that can replace these depleted resources. Therefore, it seems sensible to consider alternative approaches, like sustainable agriculture. However, many developing countries have attempted to produce sufficient food to fulfill the needs of their increasing populations. This forces them to apply more chemicals to their land because it is effective in the short term.

Most parts of India, rains are seasonal. As a consequence, ground or stored water has to be used for irrigation, industrial and domestic use. Pressure on water resources is bound to increase in the future due to population growth, urbanization, increased industrial requirement and higher living standards. To enhance the productivity of dryland agriculture, some protective irrigation will have to be provided in the areas which receive low rainfall. The water-table in many parts of India is receding and overexploitation of groundwater resources is a major threat to survival of future generations. Transgenic technologies have little to contribute towards alleviating the problems created by the overexploitation of water resources. However, if transgenic technologies can contribute towards enhancing productivity and yield stability of crops adapted to a low water requirement, the overall dependence on groundwater for irrigation will be reduced.

Comparison of Characteristics of Sustainable Agriculture and Conventional Agriculture	
Sustainable Agriculture	**Conventional Agriculture**
General	
• Long term sustainability	• Short-term benefits
• Internal solution to internal problems	• External solutions to internal problems
• Emphasis on management solutions to problems	• Emphasis on technology solutions to problems
• Responsive to feedback and participatory	• Detachment
Technical	
• Low internal input	• High external input
• To maintain soil fertility and productivity, rely upon crop rotation, recycling of residues, animal manure	• Use of synthetic compounded fertilizer
• To manage insects, weeds and other pests utilizes natural cultural and biological controls	• Use of pesticides, herbicide, growth regulators, Pharmaceuticals and livestock feed additives
• R & D emphasis on farming system and system approach	• Emphasis on individual crop
• Diversified enterprises within the farm, crops grown and cultivars used.; biodiversity	• Intensive mono cropping genetic erosion
• Emphasis is on working with natural processes	• Emphasis is on controlling natural processes
• Recognizes location specificity of technologies, use of appropriate and indigenous technologies	• Belief in universal technologies, e.g., pesticides, fertilizers, use of imported and packaged technologies
• Use of technologies that preserve and enrich the natural resource base.	• Use of technologies that exploits and destroys the natural resource base
Economic	
• Priority is food security	• Export and profit oriented
• Relies on available indigenous farm resource self reliant	• Capital intensive, usually need credit
• Places high value on human fulfillment and the environment	• Emphasis on commodity exchange in the market
Socio-political	
• Belief in accountability and value laden	• socio-politically detached and ignore consequences

Replacement of traditional mixed cropping patterns with monocultures requiring unacceptable levels of pesticide use and negative impacts on biodiversity, as well as irrigation induced problems of mineral contamination, increased salinity and the lowering of water tables. At the same time, the Green Revolution may have relieved pressure to expand agriculture into ecologically fragile areas and reduced the dependence of rural areas on unsustainable resource extraction activities.

The real problem will be the process of integration of the different technologies into the farming system. That is the process of sorting all the various legitimate factors that prevent or discourage farmers from accepting sustainable development, and adjusting the socio-economic environment to overcome these. It will have to be largely a farm level diagnostic effort to make certain the required extra labour is in fact available. There is actually enough organic residues to justify the composting process, and this could provide the desired nutrients for the entire intended area, or, if not, the percentage of the area that can be provided. Likewise, the extra land for building leguminous hedgerows needs to be available, and not already committed to subsistence cropping. In many intensive farming communities in Asia, villages are immediately adjacent to each other with all in-between land already allocated and in use. Thus, the extra land may just not be there.

The problems have increased dramatically in recent years this includes contamination of water by pesticides and fertilizers, contamination of food and fodder by residues of pesticides, nitrates and antibiotic, workers and public, disruption of ecosystem and harm of wildlife, damage to farm and natural resources by pesticides, causing workers and public, disruption of ecosystem and harm of wildlife. Considering the above problems of Indian agriculture I would like to suggest following measures to overcome the problems.

Measures for Sustainable Agricultural Development

Some of the major bacterial and fungal pathogens need to be more intensively studied, both for variability at the molecular levels and through differentials. Participation of laboratories from India in structural and functional genomics work through international collaborations should be encouraged. India did well to participate in the international rice genome sequencing effort. Such participations should continue on sequencing genomes of model legume species and some of the important pathogens of crop plants.

A major initiative will be required to attract talented students to agricultural biotechnology. At the undergraduate and postgraduate levels, curricula are outdated. These need to be changed.

Indian fertilizer industry should be given incentives to enter the business of producing and delivering quality seed of both hybrids and pure lines to the farmers. With their strong distributional networks and ties with the farmers, the fertilizer companies may be able to bring about a rapid turnaround in the seed sector.

Advisory services serve as an independent source of information and guidance, and governments can maintain databases for monitoring agriculture's use of its natural resource base and the consequences. Farmers' responsiveness to market signals and their uptake of more resource-efficient technologies depend on good dissemination of information about technological and management options and their consequences for better resource management. For example, agricultural biotechnology can contribute to improvements in resource management, low-pesticide strategies and food quality. Farmers need a good basic education so they can evaluate the information received as well as flexible skills so they can adopt new strategies. Development of transgenic for reducing post-harvest losses should be given high priority. Basic work on senescence retardation will have to be supported.

One of the objectives of integrated land use planning is to identify areas where a particular technology can provide multiple benefits. Techniques that improve soil fertility, thereby both increasing agricultural productivity and enhancing biodiversity, are a good example. Such techniques, however, must also be matched to those areas where they can be adopted and will thrive. In order to identify these opportunities, planners need relevant information about interactions between farming practices and ecosystem properties, and appropriate spatial databases.

Rainwater harvesting include groundwater recharge and improvements in soil fertility from nutrients and organic matter contained in run-off water. Moreover, water harvesting systems are thought to protect the soil from Stalinization. Enhancing local water supply in this way can be a first step towards controlling desertification, as it can make possible the reestablishment of pasture and reforestation.

Government action plans for coherent, multi-sectoral climate adaptation strategies in agriculture are needed. Farmers in developed countries can employ various strategies, such as shifting sowing and harvesting dates, adopting different varieties or species, modifying field operations such as tillage methods and fertilizer applications, adapting animal housing, and changing grain drying and storing methods. Governments will need to implement other adaptation responses, including modifying water supply and irrigation systems, stimulating the efficiency of water use, strengthening forest fire management provisions, targeting specific local vulnerabilities, increasing training and education, refocusing agricultural research priorities and strengthening extension and communication systems

It is essential to prevent climate change from derailing the development process. There are many barriers to incorporating climate change into development programmes as well as concrete steps to be taken. Raising awareness at all levels, from government down to local communities, is a precondition for ensuring that climate risk management becomes better integrated into development programmes and practices. At government level, adaptation should be approached cross-sectorally and responsibility should not be left to Environment Ministries. Government budgets should include provisions for adaptation initiatives.

Gender inequality has broader economic and social consequences than just reduced agricultural output. If women had more economic autonomy and control, fertility rates would fall, a larger proportion of household resources would be spent on nutrition, health and education, and girls would be more likely to attend school for longer. In addition, if women farmers had secure land rights and access to credit, they would be more willing to adopt medium-term soil conservation and improvement strategies and invest in micro-projects like small-scale rainwater harvesting. The triple dividend in economic, social and environmental terms of such actions is clear. With the above measures following can be implemented :

1. to incentivize the States to increase their investment in Agriculture and allied sectors,
2. to provide flexibility and autonomy to the States in planning and executing programmes for Agriculture,
3. to ensure the preparation of Agriculture plans for the Districts and the States,
4. to achieve the goal of reducing the yield gaps in important crops,
5. to maximize returns to the farmers,
6. to address the agriculture and allied sectors in an integrated manner.

Long Run Strategies for Thrust Areas

For the sustainable agriculture development only measures are not sufficient it required log run strategy of planning. This strategy brings every piece of cultivable land under cropping and taking sustained efforts to bring the fallow lands under cultivation. This required

1. Intensifying crop diversification to increase cropping intensity.
2. Ensuring standardization of quality norms for seeds and pedigree planting material especially in horticulture crops.
3. Encouraging judicious use of water through water saving devices and

latest technologies such as drip fustigation, pit method of planting in sugarcane, SRI techniques in paddy etc.

4. Paying adequate care to the soil health care to attain potential yield, suggesting the right crops to conserve natural resources and carry out sustainable agriculture and issue of soil health card to the farmers.
5. Revamping the agricultural extension so as to deliver agriculture technology to the door steps of the farmers.
6. Emphasizing promotion of Hi-tech Precision Farming, strengthening the post harvest handling practices, concentrating on processing and value addition, building linkages with the private sector at input and marketing end to improve farmer's income
7. Providing alternate livelihood option to the farmers especially dryland agriculture through farming systems approach.
8. Providing a social security net for the farmers through crop insurance.
9. To provide and produce good quality of seeds, seedlings, plants and fruit saplings at various Government Seed farm cum Horticulture Nursery for promoting the vegetable and fruit cultivation.
10. To increase the area, production and productivity of Horticultural crops need to provide the technical knowhow for growing high value crops to generate the more income by the farmers.
11. To provide training to un-employed candidate also who are interested to adopt the horticulture as profession, in public as well as to arrange vocational training for farmers.
12. To arrange the demonstration cum training for farmers, engaged in horticultural activities of commercial horticulture crops, like vegetables, fruits and ornamental plants etc.
13. To arrange the Exhibition/Kisan Fairs/show of vegetables and fruits for the competition.
14. To provide the vegetable kits (including quality seed & other inputs) to the farmers for cultivation of Bhindi, Capsicum, Brinjal, Cabbage, Cauliflower etc. an area of half an acre vegetables & off-seasons crops to increase the production & productivity of the same and to motivate the farmers.

Conclusion

Agriculture has been and will always remain the most important sector of a national and global economy. It provides the ultimate in essential food and fibre for the world's population. No industrial substitutes have yet been found to replace food requirements. Thus the long term survival of mankind will depend on the sustainability of the global agricultural economy. However, in its purest form world population has already significantly exceeded the true sustainable limits of the planet. Sustainability would therefore, have to be an exercise in optimizing humankind's need for agricultural commodities with the necessity to protect the environment so it will continue to provide the necessary biological products. This can be particularly difficult, for the smallholders of Asia and the Pacific, as they remain primarily interested in increasing productivity to meet individual and national food security requirements. Most smallholders are already operating at the optimal their limited resources will allow, and thus do not have many of the necessary extra resources to implement sustainable development practices. It will require some additional government policy adjustments to make many desired sustainable practices an economically viable alternative to farm communities.

Reference

Clive A. Edwards, Rattan Lal, Patrick Madden, Robert H. Miller, and Gar House, editors, Sustainable Agricultural Systems (Soil and Water Conservation Society, 1989). Chapters 1-6 and Chapter 38.

Food, Agriculture, Conservation, and Trade Act of 1990 (FACTA), Public Law 101-624, Title XVI, Subtitle A, Section 1603 (Government Printing Office, Washington, DC, 1990) NAL Call # KF1692.A31 1990.

Garth Youngberg and Richard Harwood, "Sustainable Farming Systems: Needs and Opportunities," American Journal of Alternative Agriculture (1989) 4(3 & 4): p. 100.

International Food Policy Research Institute (IFPRI). 1995. A 2020 vision for food, agriculture, and the environment: The vision, challenge, and recommended action. Washington, D.C.: IFPRI.

J. Patrick Madden and Scott G. Chaplowe, editors, For All Generations: Making World Agriculture More Sustainable (World Sustainable Agriculture Association, 1996). Chapter 1 and Chapter 10.

NGO Sustainable Agriculture Treaty (Global Forum at Rio de Janeiro, June 1-15, 1992). Available at Information Habitat Website: http://habitat.igc.org/treaties/at-20.htm (8/23/07).

Stephen R. Gliessman, "An Ecological Definition of Sustainable Agriculture," Principles of Agro ecology and Sustainability *(1998)*. Available at Agro ecology Home Website: http://agroecology.org/Principles_Def.html (6/9/08)

Swaminathan, M. S., Sustainable Agriculture: Towards an Evergreen Revolution, Konark Publishers, Delhi, 1996.

World Commission on Environment and Development (1987) Our Common Future, Oxford University Press, Oxford.

13

India's Sustainable Agriculture and Its Allied Sectors

Vivekanand and A. Mane

Introduction

Agriculture is of great importance to India. Although the overall contribution of agriculture to India's GDP is gradually declining, agriculture and its related activities continue to contribute in a significant manner to the overall growth of the Indian economy. Agriculture continues to be the most crucial sector of the Indian economy. The largest industries of the country like sugar, jute, textiles, food processing, milk, etc. are dependent on agriculture for their raw materials. Besides, the agriculture sector and rural areas are the biggest markets for low priced and middle priced consumer goods, including durable use items. One of the important recent concepts in agricultural development has been the concern for sustainability. This concern originated with the high tech, high input and high yielding systems of the developed world and its meaning and appropriateness to the developing world.

Sustainable agriculture is defined as," agriculture that balances the need for essential agricultural commodities such as food, fiber, etc. with the necessity of protecting the physical environment and public health, the foundation of agriculture."

Sustainable agricultural development/sustainable increase in crop production could be achieved by adopting a variety of agricultural technologies, which may be summed up as:

1. Improved crop management technology through the use of high yielding, input responsive and soil, climatic and biotic stresses - tolerant crop varieties;
2. Suitable cropping systems for different agro-ecological regions based on soil, terrain and climatic suitability;
3. Integrated nutrient management for improving soil productivity and minimization of the risk of pollution of soil, water and environment;
4. Integrated pest management for effective pests control as well as to reduce the adverse effects of pesticides on environment;
5. Soil and water conservation for controlling soil degradation and improving moisture availability.
6. Input use efficiency maximization in terms of economic return with minimal input.

Although the share of the agriculture sector in India's gross domestic product (GDP) has registered a steady decline from 36.4% in 1982-83 to 18.5% in 2006-07, agriculture is the mainstay of the Indian economy. This sector continues to support more than half-a-billion people, providing employment to 52% of the workforce.

Slow Growth in Agriculture Sector

Between 1950-51 and 2006-07, production of food grain increased at an average annual rate of 2.5% compared to population growth that averaged 2.1% during this period. As a result, India became almost self-sufficient in food grain and there were hardly any imports between 1976-77 and 2005-06. The gap between growth in the agriculture and non- agriculture sectors began to widen since 1981-82, and, more particularly, since1996-97.

The rate of growth of food grain production decelerated to 1.2% during 1990-2007, lower than the annual rate of growth of population, which averaged 1.9%. The per capita availability of cereals and pulses, therefore, witnessed a decline during this period. Consumption of cereals declined from a peak of 468 grams per capita per day in 1990-91 to 412 grams per capita per day in 2005-06 a decline of 13% during this period.

Agricultural Production in 2006-07 and 2007-08

The overall production of food grain was estimated at 217.3 million tones in 2006-07, an increase of 4.2% over 2005-06. Compared to the target set for 2006-07, it was, however, lower by 2.7 million tones (1.2%). The increase in production in 2006-07 was largely because of higher production of wheat by 6.5 million tones (9.3%) and of pulses by 0.8 million tones (6%). There was a decline in production of oilseeds (3.7 million tones, or 13%) compared to production in 2005-06.

The overall food grain production in 2007-08 is expected to fall short of the target by 2.2 million tones, though it is expected to be 10.1 million tones higher compared to the second estimates for 2006-07. Sugarcane production is estimated to exceed the target, though it will be lower than the previous year.

Over a medium term, there has generally been a shortfall in the achievement of targets for food grain, pulses and oilseeds between 2000-01 and 2006-07. The actual production of food grain, on an average, was 93% of the target. Actual production, however, was only 87.7% of the target for pulses, and 85.3% of the target for oilseeds. Production of sugarcane and cotton, however, was higher than the target in 2005-06 and 2006-07.

Irrigation

The total irrigation potential in the country has increased from 81.1 million ha in 1991-92 to 102.8 million ha in 2006-07. The potential created so far is estimated to be 73.5% of the ultimate irrigation potential. Of the total potential created, however, only 87.2 million ha (84.9%) is actually utilised. The pace of creation of additional irrigation potential came down sharply from an average of about 3% per annum between 1950-51 and 1989-90 to 1.2%, 1.7% and 1.8% per annum, respectively, during the Eighth, Ninth and Tenth Five-Year Plan periods.

The rate of growth of utilisation of the potential created declined to 1% per annum during the Ninth Five-Year Plan period and improved to 1.5% per annum during the Tenth Five-Year Plan period. The average annual rate of utilisation remained lower than the average annual addition to the irrigation potential, resulting in continuous erosion in cumulative utilisation. This amounts to inefficient use of funds and forgone income from irrigated lands.

Responding to the continuous decline in the rate of creation in irrigation potential, the central government initiated the Accelerated Irrigation Benefit Programme (AIBP) in 1996-97 to extend assistance in the completion of irrigation schemes which had remained incomplete. In 2007-08, as on January 29, 2008, Rs.3,127.5 crore had been released for the AIBP.

Capital Formation in Agriculture

Productivity increase in agriculture is considerably dependent on capital formation both from the public and private sectors. Gross capital formation (GCF) in agriculture as a proportion of total capital formation has shown a continuous decline. GCF in agriculture relative to GDP in this sector has shown an improvement from 9.6% in 2000-01 to 12.5% in 2006-07. This however needs to be raised to 16% during the Eleventh Five-Year Plan to achieve the target growth of 4% in this sector.

Credit Support to Agriculture

The Farm Credit Package announced in June 2004 stipulated, among other things, doubling the flow of institutional credit for agriculture in the ensuing three years. Credit flow to the farm sector was doubled during two years, as against the stipulated time period of three years.

To provide adequate and timely support from the banking system to farmers for their cultivation needs, including the purchase of inputs in a flexible and cost-effective manner, a Kisan Credit Card (KCC) Scheme was introduced in August 1988 for short- and medium-term loans.

Rate of Interest on Agricultural Loans

From kharif 2006-07, the Centre decided that farmers would receive crop loans up to a principal amount of Rs 3 lakh at a 7% rate of interest, and that the Government of India would provide the necessary interest subvention to NABARD and banks for this purpose. This policy has been continued in 2007-08 also. For this, the Government of India made provision of Rs 1,677 crore in Union Budget 2007-08.

Agricultural Insurance

The National Agricultural Insurance Scheme (NAIS) for crops has been implemented from the rabi 1999-2000 season with the objective of providing insurance coverage in the event of failure of any notified crops as a result of natural calamities, pests and diseases. At present, 10% subsidy in premium is available to small and marginal farmers, to be shared equally by the Centre and state governments. The scheme is implemented in 23 states and two union territories. Since the inception of the scheme and until rabi 2006-07, about 971 lakh farmers have been covered. The coverage area is 156 million ha and the sum insured is Rs 92,618 crore. Claims to the tune of around Rs 9,855 crore have become payable against the premium income of around Rs 2,943 crore, benefiting nearly 270 lakh farmers.

As announced in Union Budget 2007-08, a Weather-based Crop Insurance Scheme (WBCIS) was implemented in selected areas of Karnataka on a pilot basis. The WBCIS intends to provide insurance protection to farmers against deficit and excess rainfall. It has the advantage of being able to settle claims within the shortest possible time. Agriculture Insurance Company of India Ltd (AIC) has implemented the pilot WBCIS in Karnataka during the 2007 kharif season, covering eight rain-fed crops, insuring crops on nearly 50,000 ha for a sum insured of Rs 50 crore.

The WBCIS is being implemented in 2007-08 on a larger scale in selected areas of 12 states for the rabi 2007-08 season. In addition to AIC, private insurers like ICICI-LOMBARD General Insurance Company and IFFCO-TOKIO General Insurance Company have been included for selected areas.

Reforms Through the APMC Act

To inject dynamism and efficiency into the marketing system, large investments are required to develop a post-harvest and cold chain infrastructure close to the farmers' fields. Accordingly, the ministry of agriculture circulated a model Agriculture Produce Marketing Committee (APMC) Act, 2003, and suggested amendments to the state APMC Acts to promote investments in marketing infrastructure, motivate the corporate sector into undertaking direct marketing and facilitate a national integrated market. The ministry of agriculture formulated model APMC Rules and circulated it among the states. The ministry has requested states and union territories to complete the process of amendments and notification of rules by 2007-08.

National Food Security Mission

The department of agriculture and cooperation, ministry of agriculture, has launched a centrally-sponsored scheme — the National Food Security Mission (NFSM) — to increase the production of rice, wheat and pulses by 10, 8 and 2 million tones, respectively, over the benchmark levels of production, by the end of the Eleventh Five-Year Plan period.

Activities of the NFSM relate to:

1. Popularisation of new varieties.
2. Demonstrations of improved production technology.
3. Distribution of quality high-yielding variety (HYVs) seeds and hybrids.
4. Training and mass media campaigns, including awards for the best performing districts.
5. Support for micronutrients.

The NFSM is being implemented in 305 districts of 16 states in the country. An amount of Rs 149.4 crore has been released to the states to plan and implement various interventions during 2007-08.

Rashtriya Krishi Vikas Yojana (RKVY)

To offer incentives to states to increase the share of investments in agriculture in their plans, a Rashtriya Krishi Vikas Yojana (RKVY) with an allocation of Rs 25,000 crore has been proposed for the Eleventh Five-Year Plan.

The RKVY will be a state plan scheme and eligibility for assistance under it will depend on the amount provided in state budgets for agriculture and allied sectors, over and above the baseline percentage expenditure incurred on agriculture and allied sectors. Funds under the RKVY will be provided to the states as a 100% grant by the central government. An outlay of Rs 1,500 crore has been approved for 2007-08.

Strategies for Sustainable Agriculture

The most effective strategies to do this include:

1. **Strengthening Panchayati Raj**: The voice of the poor is best expressed through Panchayati Raj institutions. The poor need to be made aware of their power to exercise democratic control over local representatives and public services.
2. **Empowering women**: Encouraging women to make aware of their rights and capacities, particularly the importance of their participation in local governance, income-generation and decision-making.
3. **Reducing social exclusion**: To increase awareness of rights and catalyse change in institutions and policies to reduce discrimination against tribals and dalits, the landless, women, children and minorities.
4. **Advocating policy changes**: The absence of relevant and effective government policies compounds the problems of the poor. In other instances, pro-poor policies exist but are not implemented. So, to avoid, there should be necessary work on advocacy efforts for the formulation and implementation of policies in favour of the poor at the state and national level.
5. **Generating livelihood opportunities**: For the poor to work towards the long-term outcomes listed above, it is essential that their immediate, basic needs be met. Therefore, it is necessary to establish self-help groups (SHGs). SHGs also provide a platform for addressing other key issues.

(a) Awareness generation through meetings, rallies and camps

(b) Initiating and building the capacity of community-based organisations through training programmes, exposure visits and promotion of income-generation activities

(c) Communication through media like wall paintings, posters, puppet shows and newsletters

(d) Advocacy through public hearings, media exposure and protests

SUSTAINABILITY IN ALLIED SECTORS

(A) Indian Horticulture Sector

The horticulture sector encompasses a wide range of crops namely fruits, vegetables, potato, tubers, ornamental, medicinal and aromatic, spices and plantation crops. In recent past, horticulture has emerged as an indispensable part of agriculture, offering a wide range of choices to the farmers for crop diversification. Diversification to horticulture crops is now the major option to improve livelihood security of small farmers, improved employment opportunities to attain food and nutrition security, improve income through value addition and foreign exchange through exports. It also provides ample opportunities for sustaining large number of agro-industries, which generate substantial employment opportunities.

The horticulture sector contributes around 28% of the GDP from about 13.08% of the area and 37% of the total exports of agricultural commodities. During the previous three plan periods, focused attention was given to horticultural research and development which placed India as the second largest producer of fruits and vegetables. largest producer and consumer of cashew nut, tea spices, third largest producer of coconut, forth largest producer and consumer of rubber and sixth largest producer of coffee in the world. Today, as a result synergy between focused research, technological and policy initiatives, high efficiency inputs, horticulture in India, has become a sustainable and viable venture for the small and marginal farmers. Besides, it has also started attracting entrepreneurs for talking up horticulture as a commercial venture.

Scope and Importance

Horticultural crops play a unique role in India's economy by improving the income of the rural people. Cultivation of these crops is labour intensive and as such they generate lot of employment opportunities for the rural population. Fruits and vegetables are also rich source of vitamins, minerals, proteins and carbohydrates etc. which are essential in human nutrition. Hence, these are referred to as protective foods and assumed great importance as nutritional security of the people. India with diverse soil and climate comprising several agro-ecological regions provides ample opportunity to grow a variety of horticulture crops. It is estimated that all the horticulture crops put together cover nearly 11-6 million hectares area with an annual production of 91 million tones. Though these crops occupy hardly 7% of the cropped area they contribute over 18% to the gross agricultural output in the country.

Fruits and vegetables are not only used for domestic consumption and processing into various products (Pickles, preserves sauces, jam, jelly sques, etc.) but also substantial quantities are exported in fresh and processed form, bringing much-needed foreign exchange for the country. These groups of crops also provide ample scope for achieving bio-diversity and diversification to maintain ecological balance and to create sustainable agriculture and can make an impact on the national economy in the years to come.

It is estimated that India has 240 million acres of cultivable wasteland, which is lying idle, which can be brought under orchard crops without curtailing the area under food crops. The country has abundant sunshine throughout year, a surplus labour and widely varied agro-climatic condition, which offers high potential for successful and profitable commercial horticulture.

3rd Indian Horticulture Congress 2008

The Horticulture Society of India is organized 3rd Indian Horticulture Congress 2008 with the theme of New R & D Initiatives in Horticulture for Accelerated Growth and Prosperity in collaboration with Ministry of Agriculture, Government of India, Orissa University of Agriculture & Technology and Orissa Horticulture Society. A number of distinguished invited speakers will provide an overview of topics related to the new initiatives in horticulture R & D for Accelerated Growth and Prosperity. The major objective of this Congress is to provide a forum for the exchange of ideas among researchers, academicians, policy makers, development workers and industry professionals etc involved in growth of horticulture sector.

ISAP

Indian Society of Agribusiness Professionals (ISAP), a non-profit organization has a network of agriculture and allied sector professionals in India and developing countries. It is a growing network encompassing over 15,000 registered associate members, comprising 1500 Agri-Experts, 525Partner NGOs, over 824 Individual users and more than 1050 researchers. ISAP has already established its presence in 12 states in India for the development of farm sector. ISAP has successfully carried out several assignments for the development of horticulture sector, supported by government of India, state governments, national and international organizations and also organized various event including buyer seller meets, conferences, training programmes etc.

Horti EXPO 2008

Coinciding with the 3rd Indian Horticulture Congress 2008, the Indian Society of Agribusiness Professional (ISAP) New Delhi is organizing a mega Horti Food Technology Exhibition HORTI EXPO2008 from 6th to 9th November 2008 at OUAT Ground Bhubaneshwer (at the venue of the Congress).

The objectives of the Expo are to showcase the products and services related to horticultural diversity inputs, productivity and post harvest management aspects besides marketing and trade related issues. It would have the participation of central and state government departments, research and development institutions, boards, with the producers, processors, corporates, technology providers, exporters, equipments & machinery manufactures and consultants etc.

Participating Sectors

1. Agro and Rural Industry
2. Commodities Tea, Coffee, Coconut, Coir
3. Micro Irrigation and Plasticulture Industry
4. Packaging and Processing Industry
5. Banks and Financial Institutions
6. Equipments & Machinery Manufacturers
7. Cold Chain and Food Processors
8. Retail Industry & bulk buyers
9. Bio Technology
10. Processed Food Products
11. Research and Technology

(b) Indian Sericulture Sector

Sericulture is one of the rural based agro industry with global reach. While providing sustainable income and employment opportunities to the rural poor who are the main practitioners, silk production activity fetches annual export earnings of more than US$600 million. Some unique features of the silk sector are its rural nature, agro based, ecologically and economically sustainable activity for the poor, small and marginal farmers, agriculture labour and women in particular. Many studies indicated that 60% of the activities in the pre-cocoon and post-cocoon sectors are carried out by women.

Sericulture is a labour intensive industry in all its phases. It can generate employment upto 11 persons for every kg of raw silk produced. Out of which more than 6 persons are women. More than 60.00 lakh persons are employed as full time workers in the production chain out of which 35-40 lakh persons are women.

Though India produces all four varieties of silk, mulberry sericulture dominates with 89% in total production and 95% in exports. 98% of mulberry silk production takes place in the states of Andhra Pradesh, Karnataka, Tamilnadu, West Bengal and Jammu & Kashmir. Central Silk Board and the State departments had adequate experience of implementing the women oriented programmes.

Sericulture Activities

Moriculture

Cultivation of mulberry plants is referred to as Moriculture. It is an agricultural activity. In Tamil Nadu, Andra Pradesh & Karnataka mulberry cultivation is mainly taken up in irrigated condition. Flat, deep, fertile, well drained loamy and clay loamy with good moisture holding capacity soil is ideal for mulberry cultivation.

Silkworm Seed Production

The silkworm seed production centers are referred to as grainages. The silkworm seed known as Disease Free Layings are prepared in their centers and supplied to the farmers for rearing. Both Government and private sector grainages are involved in this activity.

Silkworm Rearing

Silkworm Rearing is considered to be an agro based cottage industry since it involves mulberry cultivation. Silkworms are reared for the production of "cocoons" which is the raw material for silk production. The farmers rear silkworms and produce cocoons. By marketing the cocoons the farmers earn money. It is ideally suited for the rural areas of Sericulture States. Silkworms are reared in well ventilated rearing shed following shoot rearing method.

Marketing of Cocoons

The farmers can sell the cocoon produced by them in the nearest Govt. Cocoon Markets. In the cocoon markets reasonable floor price is fixed by scientific methods and the final selling price is decided in the open auction. Here, silk reelers buy the cocoons produced by the farmers for producing silk.

Silk Reeling

Extraction of silk filament from cocoons by employing a set of processes is known as silk reeling. Presently silk reeling is done using three types of reeling devices viz. Charka, Cottage basins and Multi-end basins. Poor quality cocoons can be reeled economically on charka. About 50% of silk produced is of charka and about 35-40% is at cottage basins and a small quantity of silk is from multiend reeling.

Marketing of Raw Silk

The raw silks produced by the silk reelers are marketed directly to the weavers or through Silk Exchanges functioning in the Sericultural States.

In the Silk Exchanges all the raw silk lots brought by the silk reelers are tested for quality. On the basis of quality of raw silk and the price prevailed in major marketing centre, floor price is fixed and then they are auctioned. The Silk Exchange enables the reelers to get immediate cash for the raw silk transacted by them.

Silk Weaving

The raw silk cannot be directly used for weaving. The raw silk is to be twisted before they are fed into looms. The operation of conversion of raw silk into twisted silk, is termed as twisting. The twisted silk is referred to as Ready Silk. Twisting is undertaken either by separate entrepreneurs or by the weavers themselves. The silk weaving is done either on handlooms or power looms. The traditional silk sarees and dhoties are made on handlooms whereas the printed sarees, dress materials, etc., are made on power looms.

Sericulture in India

In India, Sericulture is essentially a village-based industry providing employment to a sizable section of the population. Although Sericulture is considered as a subsidiary occupation, technological innovation has made it possible to take it up on an intensive scale capable of generating adequate income. It is also capable of providing continuous income to farmers.

India is the second largest producer of silk in the world with an annual silk production of around 16,500 M.Tons. All the known varieties of silk, viz. Mulberry, Eri, Muga and Tasar are produced in India. Mulberry silk is the most popular variety in India, contributing more than 87% of the Country's silk production. Silk and silk goods are very good foreign exchange earners. Export potential of this sector is promising as silk production in Japan is declining and that of China, the largest silk production in the World, it is stagnant or declining.

In India, because of the prevalence of favourable climatic conditions, mulberry is cultivated mainly in five states, viz., Karnataka, Andhra Pradesh, Tamil Nadu, West Bengal and Jammu & Kashmir. These five states collectively account for 97% of the total area under mulberry cultivation and 95% of raw silk production in the country.

Central Silk Board

For the development of silk industry in India, the Central Silk Board, a statutory body, is functioning under the administrative control of the Ministry of Textiles, Govt. of India, with its Headquarters at Bangalore.

The following are the important functions assigned to the Board.

1. Promoting the development of silk industry by such measures as it thinks fit.
2. Undertaking, assisting and encouraging scientific, technological and Economic Research.
3. Devising means for improved methods of mulberry cultivation, silkworm rearing, developing and distributing healthy silkworm seeds, improving methods of silk reeling, improving the quality and production of raw silk.
4. Improving the marketing of raw silk.
5. The collection and compilation of statistics.
6. Advising the Govt. of India an all matters relating to the development of silk industry including import and export of raw silk.

Conclusions

1. The opportunities for agricultural development in low-potential areas are more limited, although there appears to be good potential to increase the overall productivity of land use through better management of grazing lands and wastelands, particularly in lower population density areas where substantial areas of such lands are still present.

2. Research and extension programs will need to take a broader focus, emphasizing market opportunities for new commodities, management of animal health, integrated pest management, and integrated soil nutrient management.
3. Assuring adequate provision of inputs and credit, and development of the marketing system are critical to all commercial strategies. Development of processing facilities and marketing institutions (such as cooperatives and contract farming), facilitated by a supportive policy environment, are needed.
4. There should be priority for road development in high-potential areas to the rural markets, especially where population density is high. There is good potential for intensified production of high-value perennial crops in these areas if roads are adequate. However, achieving this potential first requires assuring food security, which is likely to be most economical by increasing productivity in food crop production.
5. For low-potential areas without good potential for irrigation (especially with lower population density), priority should be placed on promoting increased productivity of all land, including grazing lands and wastelands. Cautious efforts are needed by governments and NGOs to catalyze development of local institutions in order to intensify management of grazing lands.
6. Addressing problems of poverty, low agricultural productivity and resource degradation will therefore require strategies that address the needs of the poor as well as the more well-endowed.
7. The present system is not sustainable because it is not resources efficient, environmentally sound and socially viable. Failure to fulfil the increasing demands for food and fuel is the main reason for this lack of sustainability.

References

Agricultural Research Data Book (2002), IASRI, New Delhi.

Agricultural Statistics at a Glance (2001), Directorate of Economics and Statistics, Department of Agriculture and Cooperation, Ministry of Agriculture, Govt. of India.

A Brief Note On Sericulture, By Thiru. Harmander Singh., I.A.S.

http://www.empowerpoor.com/backgrounder.asp?report=726

Indian Horticulture Data Base (2001), National Horticulture Board, Ministry of Agriculture, GOI.

Mr Vedprakash Sharma, Event Coordinator, Indian Society of Agribusiness Professionals (ISAP) R 289 A, Greater Kailash- 1, New Delhi 110 048 Website: www.isapindia.org

Predominent cropping systems of India: Technologies and strategies by Yadav, A.K., Prasad Kamta and Singh, A.K.(1998).

Reardon, T., E. Crawford, and V. Kelly. 1996. Links between nonfarm income and farm investment in African households: Adding the capital market perspective.*American Journal of Agricultural Economics* 76: 1172-1176.

Swaminathan, M. S., Sustainable Agriculture: Towards an Evergreen Revolution, Konark Publishers, Delhi, 1996.

14

Environment and Sustainable Agricultural Development

P. Loganathan and A. Sakthivel

Introduction

Sustainable development is a pattern of resource use that aims to meet human needs while preserving the environment so that these needs can be met not only in the present, but in the indefinite future. It "meets the needs of the present without compromising the ability of future generations to meet their own needs." Sustainable development is the way in which developing nations undergoing the process of industrialization will avoid becoming like current industrialized carbon intensive nations with high level of emissions. The field of sustainable development can be conceptually broken into three constituent parts: environmental sustainability, economic sustainability and sociopolitical sustainability.

Scope and Definitions

The concept has included notions of weak sustainability, strong sustainability and deep ecology. Sustainable development does not focus solely on environmental issues.

The United Nations 2005 World Summit Outcome Document refers to the "interdependent and mutually reinforcing pillars" of sustainable development as economic development, social development, and environmental protection.

Economic Sustainability

Identified information, integration, and participation as key building blocks to help countries achieve development that recognizes these interdependent pillars. It emphasizes that in sustainable development everyone is a user and provider of information. It stresses the need to change from old sector-centered ways of doing business to new approaches that involve cross-sectoral co-ordination and the integration of environmental and social concerns into all development processes.

According to Hasna, sustainability is a process which tells of a development of all aspects of human life affecting sustenance. It means resolving the conflict between the various competing goals, and involves the simultaneous pursuit of economic prosperity, environmental quality and social equity famously known as three dimensions with is the resultant vector being technology, hence it is a continually evolving process;

Green Development

Development is generally differentiated from sustainable development in that Green development prioritizes what its proponents consider to be environmental sustainability over economic and cultural considerations. Proponents of Sustainable Development argue that it provides a context in which to improve overall sustainability where cutting edge Green development is unattainable. For example, a cutting edge treatment plant with extremely high maintenance costs may not be sustainable in regions of the world with fewer financial resources. An environmentally ideal plant that is shut down due to bankruptcy is obviously less sustainable than one that is maintainable by the community, even if it is somewhat less effective from an environmental standpoint.

Sustainable Agriculture

The word "sustain," from the Latin sustinere (sus-, from below and tenere, to hold), to keep in existence or maintain, implies long-term support or permanence. As it pertains to agriculture, sustainable describes farming systems that are "capable of maintaining their productivity and usefulness to society indefinitely. Such systems... must be resource-conserving, socially supportive, commercially competitive, and environmentally sound.

The term sustainable agriculture means an integrated system of plant and animal production practices having a site-specific application that will, over the long term:

1. satisfy human food and fiber needs;
2. enhance environmental quality and the natural resource base upon which the agricultural economy depends;

3. make the most efficient use of nonrenewable resources and on-farm resources and integrate, where appropriate, natural biological cycles and controls;
4. sustain the economic viability of farm operations; and
5. enhance the quality of life for farmers and society as a whole."

Sustainable agriculture: A whole-systems approach to food, feed and other fiber production that balances environmental soundness, social equity, and economic viability among all sectors of the public, including international and intergenerational peoples. Inherent in this definition is the idea that sustainability must be extended not only globally, but indefinitely in time, and to all living organisms including humans.

Sustainable Agro Ecosystems

1. maintain their natural resource base,
2. rely on minimum artificial inputs from outside the farm system,
3. manage pests and diseases through internal regulating mechanisms,
4. recover from the disturbances caused by cultivation and harvest."

Sustainable agriculture does not refer to a prescribed set of practices. Instead, it challenges producers to think about the long-term implications of practices and the broad interactions and dynamics of agricultural systems. It also invites consumers to get more involved in agriculture by learning more about and becoming active participants in their food systems. A key goal is to understand agriculture from an ecological perspective—in terms of nutrient and energy dynamics, and interactions among plants, animals, insects and other organisms in agroecosystems—then balance it with profit, community and consumer needs."

Sustainable agriculture is "a way of practicing agriculture which seeks to optimize skills and technology to achieve long-term stability of the agricultural enterprise, environmental protection, and consumer safety. It is achieved through management strategies which help the producer select hybrids and varieties, soil conserving cultural practices, soil fertility programs, and pest management programs. The goal of sustainable agriculture is to minimize adverse impacts to the immediate and off-farm environments while providing a sustained level of production and profit. Sound resource conservation is an integral part of the means to achieve sustainable agriculture."

Sustainable Farming Practices Commonly Include

1. crop rotations that mitigate weeds, disease, insect and other pest problems;

provide alternative sources of soil nitrogen; reduce soil erosion; and reduce risk of water contamination by agricultural chemicals

2. pest control strategies that are not harmful to natural systems, farmers, their neighbors, or consumers. This includes integrated pest management techniques that reduce the need for pesticides by practices such as scouting, use of resistant cultivars, timing of planting, and biological pest controls
3. increased mechanical/biological weed control; more soil and water conservation practices; and strategic use of animal and green manures
4. use of natural or synthetic inputs in a way that poses no significant hazard to man, animals, or the environment.

This approach encompasses the whole farm, relying on the expertise of farmers, interdisciplinary teams of scientists, and specialists from the public and private sectors.

1. "Sustainable agriculture is a model of social and economic organization based on an equitable and participatory vision of development which recognizes the environment and natural resources as the foundation of economic activity. Agriculture is sustainable when it is ecologically sound, economically viable, socially just, culturally appropriate and based on a holistic scientific approach.
2. "Sustainable agriculture preserves biodiversity, maintains soil fertility and water purity, conserves and improves the chemical, physical and biological qualities of the soil, recycles natural resources and conserves energy. Sustainable agriculture produces diverse forms of high quality foods, fibers and medicines.
3. "Sustainable agriculture uses locally available renewable resources, appropriate and affordable technologies and minimizes the use of external and purchased inputs, thereby increasing local independence and self sufficiency and insuring a source of stable income for peasants, family and small farmers and rural communities. This allows more people to stay on the land, strengthens rural communities and integrates humans with their environment.

Sustainable agriculture respects the ecological principles of diversity and interdependence and uses the insights of modern science to improve rather than displace the traditional wisdom accumulated over centuries by innumerable farmers around the world.

Industrial Farming

The long-term viability of our current food production system is being questioned for many reasons. The news media regularly present us with the paradox of starvation amidst plenty—including pictures of hungry children with supermarket ads. Possible adverse environmental impacts of agriculture and increased incidence of food borne illness also demand our attention. "Farm crises" seem to recur with regularity.

The prevailing agricultural system, variously called "conventional farming," "modern agriculture," or "industrial farming" has delivered tremendous gains in productivity and efficiency. Food production worldwide has risen in the past 50 years; the World Bank estimates that between 70 percent and 90 percent of the recent increases in food production is the result of conventional agriculture rather than greater acreage under cultivation. U.S. consumers have come to expect abundant and inexpensive food.

Conventional farming systems vary from farm to farm and from country to country. However, they share many characteristics: rapid technological innovation; large capital investments in order to apply production and management technology; large-scale farms; single crops/row crops grown continuously over many seasons; uniform high-yield hybrid crops; extensive use of pesticides, fertilizers, and external energy inputs; high labor efficiency; and dependency on agribusiness. In the case of livestock, most production comes from confined, concentrated systems.

Philosophical underpinnings of industrial agriculture include assumptions that (a) nature is a competitor to be overcome; (b) progress requires unending evolution of larger farms and depopulation of farm communities; (c) progress is measured primarily by increased material consumption; (d) efficiency is measured by looking at the bottom line; and (e) science is an unbiased enterprise driven by natural forces to produce social good.

Ecological Concerns

Agriculture profoundly affects many ecological systems. Negative effects of current practices include the following:

1. Decline in soil productivity can be due to wind and water erosion of exposed topsoil; soil compaction; loss of soil organic matter, water holding capacity, and biological activity; and salinization of soils and irrigation water in irrigated farming areas. Desertification due to overgrazing is a growing problem, especially in parts of Africa.
2. Agriculture is the largest single non-point source of water pollutants

including sediments, salts, fertilizers (nitrates and phosphorus), pesticides, and manures. Pesticides from every chemical class have been detected in groundwater and are commonly found in groundwater beneath agricultural areas; they are widespread in the nation's surface waters. Eutrophication and "dead zones" due to nutrient runoff affect many rivers, lakes, and oceans. Reduced water quality impacts agricultural production, drinking water supplies, and fishery production.

3. Water scarcity in many places is due to overuse of surface and ground water for irrigation with little concern for the natural cycle that maintains stable water availability.
4. Other environmental ills include over 400 insects and mite pests and more than 70 fungal pathogens that have become resistant to one or more pesticides; stresses on pollinator and other beneficial species through pesticide use; loss of wetlands and wildlife habitat; and reduced genetic diversity due to reliance on genetic uniformity in most crops and livestock breeds.
5. Agriculture's link to global climate change is just beginning to be appreciated. Destruction of tropical forests and other native vegetation for agricultural production has a role in elevated levels of carbon dioxide and other greenhouse gases. Recent studies have found that soils may be sources or sinks for greenhouse gases.

Economic and Social Concerns

Economic and social problems associated with agriculture can not be separated from external economic and social pressures. As barriers to a sustainable and equitable food supply system, however, the problems may be described in the following way:

Economic pressures have led to a tremendous loss of farms, particularly small farms, and farmers during the past few decades—more than 155,000 farms were lost from 1987 to 1997. This contributes to the disintegration of rural communities and localized marketing systems. Economically, it is very difficult for potential farmers to enter the business today. Productive farm land also has been pressured by urban and suburban sprawl—since 1970, over 30 million acres have been lost to development.

Conclusion

Many in the agricultural community have adopted the sense of urgency and direction pointed to by the sustainable agriculture concept. Lack of sharp definition has not lessened its authenticity. Sustainability has become an integral component of many government, commercial, and non-profit agriculture research efforts, and it is beginning to be woven into agricultural policy. Increasing numbers of farmers and ranchers have embarked on their own paths to sustainability, incorporating integrated and innovative approaches into their own enterprises. This just-do-it attitude is the real force carrying the issue of sustainability into the next century. "The best way to communicate the meaning of sustainable agriculture is through real-life stories of farmers who are developing sustainable farming systems on their own farms. Critical discussion of the sustainable agriculture concept will and should continue. Understanding will deepen; answers will continue to come. On-going dialog is important for another reason: with more parties, each with its own agenda, jumping into the sustainable agriculture "tent," only a continued focus on the real issues and goals will keep sustainable agriculture from becoming so all-encompassing as to become meaningless.

References

John Ikerd, as quoted by Richard Duesterhaus in "Sustainability's Promise," *Journal of Soil and Water Conservation,* Jan.-Feb. 1990.

Committee on the Role of Alternative Farming Methods in Modern Production Agriculture, National Research Council, *Alternative Agriculture* (National Academy Press, 1989). Chapter 2.

Raymond Forney, *A Common Vision: Evaluating the Farming Industry's Progress Toward Sustainability,* 1998.

David Pimentel et.al., "Environmental and Economic Costs of Pesticide Use," in *BioScience,*pp.750-760.

http://www.sarep.ucdavis.edu/concept.htm

http://eap.mcgill.ca/publications/EAP34.htm

Paul F. O'Connell, "Sustainable Agriculture, a Valid Alternative," *Outlook on Agriculture* (1992)

From *NGO Sustainable Agriculture Treaty,* Global Forum at Rio de Janeiro, June 1-15, 1992:

Stephen R. Gliessman, "An Ecological Definition of Sustainable Agriculture," *Principles of Agroecology and Sustainability* (1998).

www.finminindia.nic.in

15

Health Status of Women Workers in Plantation Industry in a Select Area

Sherly Kurien

Introduction

The contribution of the women to development is seriously underestimated and socially unrecognized. The value of women's unpaid housework and community work is estimated at between 10 per cent and 35 per cent of the GDP Worldwide, and in 1996 it amounted to 12 trillion and in developing countries the women make 31 per cent of the official labour force and produce more than 55 per cent of all food grain. But still we see that the role of women in building a just and developed society through their remunerated and unremunerated work in almost all countries of the globe has been seriously underestimated (Singh, 2005).

Women constitute nearly half of the world population, receive two-thirds of the world's income and own less than a hundredth of the world's property. In India, 90 per cent of women are employed in the unorganized sector, without fair wages and occupational amenities.

Indian women, constituting nearly half of the total population, play a crucial role in the domestic sphere and also in the rural field, particularly in the agricultural sector. Yet, our traditional attitude is to treat women as second-class citizens. This is so in spite of the fact that women have proved their mettle in every walk of life. Their contribution is not given due credit. According to an ILO estimate, the value of unpaid household work constitutes 25 to 39 per cent of the GNP.

The largest number of working women in India is engaged in farming operations either as cultivators or as agricultural labours. They take up a wide variety of activities like sowing of seeds, transplanting, weeding, harvesting, preparation of compost and manure pits, application of manures, storage of seeds and food grains. An active farm women spends eight to nine hours on farm during the peak agricultural season.

Conditions of Women Workers

In the developing world, women are responsible for 50 per cent of total food production. In Tanzantu, women work an average of 2600 hours a year in agriculture, while it is only 1800 hours a year for men. In Africa as a whole, 60 per cent of all agricultural work, 50 per cent of animal husbandry, and 100 per cent of food processing is done by women. Millions of women in the third world cook and clean, sew and wash, plant and weed, care for the old and bring up the young and their working about 16 hour a day is not uncommon. The enormous work coupled with inadequate and unbalanced diet results in poor health and subsequent illness. The traditional notion of women being confined to the home, without exposure to the outside world has resulted in being deprived of proper health and other facilities. In the light of these facts an attempt was made to study the status of women in plantation work in a selected area.

The area selected for the study was Hosatti in Kilkotagiri. Kothagiri derives its name from the traditional habitat of tribal community Kotas. In fact Kotagiri was a small town which expanded because of plantation activities. Kotagiri is a town well connected both the with district headquarters and other important places within the district and with the nearest town in foothills, and about the same distance from the district headquarters Ootacamund. The northern part of taluk is covered with temperate forests and the rest of the area is covered mostly by tea plantations and vegetable.

The major plantation crop is tea which is grown extensively throughout the region. Tea gardens, small and large, lay enchantingly spread over the steep hills and deep valleys in the Nilgiris. The tea plantations here account for about 10 per cent of the total tea production in South India. In these tea gardens, also known as tea estates, several thousands of women are employed to pick tender leaves from the bushes. They traverse acres and acres of tea fields, day in and day out, to selectively pick only two leaves and a bud and collect the harvest either in jute bags or baskets slung over their back. The nimble fingers of the tea pickers work like machines and make no error about the type of leaves required for the manufacture of quality and flavor-rich tea.

Hosatti is a small village surrounded by hills and the weather is chill throughout the year which allows tea plantation activity to flourish. There are nearly 100 households and for the purpose of the study the investigator selected only 60 women who were actively engaged in plantation activity. The important characteristics of these households are that they have small holdings and their meager incomes are derived from the casual employment in tea plantation.

Socio-economic Background

The general socio-economic background of the selected women is depicted in the following section. All the households had a small piece of land to be called as their own or in the words of Amartya Sen they had an entitlement of land. The land area was very meager, to the tune of ½ acre to 1 acre. In this small plot they grow tea. In their tea estate the women work for 2-3 days a week and the rest of the days they cultivate various vegetables like carrot, beetroot, potato, cabbage etc. in a small measure. Many men in this area are not seriously involved in any work and they just while away their time in the nearby shops or community hall or in market place. The bread winners are the women. They work hard and earn a living and sustain the family. Bases on the data collected we have classified women on the basis of their income as follows and this is depicted in Table 1.

Table 1

Income-wise classification of women

Income/month (In Rs.)	Number	Per centage
Less than 5,000	20	33.3
5,001-10,000	35	58.3
Above 10,000	5	8.3

More than half of the sample size are in the income group which ranges from Rs. 5,001- 10,000 per month. Only eight per cent of the sample population had more than Rs. 10,000 as their monthly income. This reveals that all the families come below the poverty line.

Health Status

The current study probed into the health status of these women. It was revealed that nearby 90% of the women have skin diseases. This may be due to the fact that, day in and day out these women are exposed to shine and rain for their work and they work inside the tea estates or in very close contact with these plants. The other serious illness found among these women were asthma and TB. Long hours of work in the chill weather affect the lungs in the long run. All the households have the traditional way of cooking, i.e., by using firewood, they cook. The smoke which emits out of kitchen is also one reason which affects the lungs. This coupled with the climate and improper diet leads to all kinds of lung infection and finally resulting in TB. In the initial stages of asthma and TB they do not realize the seriousness they casually take it as a cold or fever and with the help of domestic medication it will be arrested temporarily. But in due course of time it appears again and again and the person loses her immunity and is prone to such serious illness. These women do not give due importance of health and this is the reason they do not approach a medical practitioner right in the beginning nor do they go to a hospital. As a last resort only they go to a Doctor and by then it becomes too serious to be treated. Hence the poor health status of the women.

The staple food item is rice and all of them take tea or coffee so frequently and this may be a reason why again whooping majority of the women are found to have diabetics. Since this is also not detected early, they land in serious illness. A disease called a sickle cell is now-a-days found among these plantation workers. It is nothing but reduction of White Blood Cells. This is mainly due to nutritional deprivation or this is the state of being anaemic. The prevalence of anaemic among women poses risk to women at various stages of her life. She will be easily susceptible to diseases and reduces the energy to do her daily chores.

Poverty condition also prevents them from meeting or consulting a doctor. If they do not work there is food in the family. Even if the husband's work he hardly brings anything to home. They are addicted to drinks and this takes away a major share of their income. Hence women have to work and definitely take care of the family. As long as they can, they continue to work and this leads to poor health. As Sathyamala (2006) points out a chronically starving population is highly susceptible to disease but is unlikely to seek early and/or adequate treatment because of cost incurred. For people who make a living by selling their physical labour on a daily basis, the high indirect cost of treatment in terms of loss of wages acts as an important determinant in medical care seeking behavior. In rural areas there is the added factor of distance and transport cots. The poor have little or no surplus to spend in terms of money, food or body stores.

The identification by the World Health Organisation (WHO) of poverty as the greatest single killer is less platitudinous than it sounds. India's mortality and morbidity profile remains overwhelmingly, dominated by poverty related infectious diseases, malnutrition and, hunger. It is well established that the calorie deprivation in rural areas has increased in the period between 1991-92 and 2002-2003. The absorption of food grains per head has fallen sharply to an abysmally low average annual level of 151 kg. By the pre-drought year 2000-01 (Patnaik, 2004). More than half the women between 15 and 49 years of age and three-fourths of the children are anaemic. Severe and chronic under-nutrition affects 18 per cent of the children. Close to 2 million children and 1,10,000 women die every year, the latter from causes related to pregnancy and childbirth. In rural India, one woman dies every five minutes while giving birth. In addition, communicable diseases continue to fester like lesions on the body of the public health structure. Together, they are responsible for the death of 2.5 million children and an equal number of adults annually (Prasad, 2006).

The frequency of incidence of common illness and the course of treatment was examined and is presented in the following table:

Table 2

Incidence of Sickness

Sickness	(Frequency (Number reporting)			
	Regularly	Once in a Month	Once in 6 Months	Rarely
Fever	25	5	7	10
Cold	35	5	-	20
Backpain	40	20	-	-
Headache	25	30	5	-
Viral fever	-	5	10	25

The major health problem that affects these women seems to be back pain. This is because they always bend down to pluck the tea leaves. Further there is bag resting at their back in which the leaves are collected. Further the exposure to the cold climate often troubles them with fever, cold and head ache. It was noticed that 50 per cent of the women go to private hospitals for treatment while 30 per cent receive medical help from government hospital and the rest of the group still takes domestic medicine. The older generation believes in the traditional treatment. Many elderly women grows medicinal plants and herbs which are very effective and they do use it regularly. But the younger women prefer to go to private doctors for fast recovery.

The cost of treatment, was reported to be Rs.100-200 for 30 women while it was more than 200 for 10 women, for a month. Though in absolute terms it looks a trivial amount, for the poor women it is a big sum of money. This is one reason for refraining themselves from attending a medical practitioner. Also a day's labour is lost. In the case of delivery, many women now seek medical help from trained health staff or admit themselves in hospital. But there were a handful that still believed in traditional system of health care delivery and would not avail outside assistance. They do not want to incur any extra cost of delivery in an institution when the birth can be conducted at home at minimal expense. This is in tune with a study done by Ramachandran (2008) on women's health in Kanyakumari District.

Health is acknowledged to be a complex issue which has strong linkages with nutritional intake, environmental sanitation, availability of safe drinking water, education and awareness and the access to preventive and curative health care. These variables were analysed for the plantation women workers too. The nutritional intake was found to be highly inadequate. They start the work very early in the morning. They go to the estate and start plucking tea leaves very early, without taking any food. This is continued till afternoon and then they take lunch. But this is an unbalanced diet. Solar also is the case with dinner. They consume lots of tea in between the work and this to a great extent nullifies their appetite.

Environmental sanitation is not very satisfactory. Personal hygiene also needs improvement. They refrain from having daily bath due to the chill weather. Drinking water is a telling indicator of good health. In this village there are two wells - one for drinking water and another one for other purposes. All the households get water from this common well. The drinking water is consumed directly and 35 women do not boil it. It seems to be safe as the well is cleaned periodically.

Health problems of plantation women are mostly work-related. Many illness or health problems arise out of infections, exposure to chill weather, psychological stresses, poor environment lack of personal hygiene and sanitation and high rates of malnutrition.

Steps towards Better Health

Surviving through a normal life cycle with good health, helps one to lead a happy and productive life. Human beings are the most productive factor of any economy. This factor needs to be healthy so as to be fully exploited to churn out much wealth in an economy. Improvement in the health status of the people has been one of the major thrust areas for the social development. The dream of healthy India can only be achieved when women's health is taken care and given priority. A holistic approach to women's health which includes nutrition, environment, sanitation and health services should be adopted. In the case of plantation women worker's special efforts should be taken to protect themselves from extreme weather conditions which would affect their health. It is a sorry state that when the price of tea goes down these poor women are at a loss. Government should give them incentive or some relief measures should be initiated to iron out such eventualities.

Education is the most powerful and potent factor for changing women's position. Most of the plantation workers have had only basic education and some are illiterates. They are not aware of their rights. Many of the Government programmes do not reach them. Women should be educated to fight for their rights. Education on nutrition and hygiene is very important as this forms the core of health. Steps should be taken to remove drudgery of women both at home and work place. Provision of health services which are easily accessible to women at affordable prices should be promoted.

Conclusion

The contribution of women to economic development is enormous and this is more so in the case of working women in unorganized sector. There is a growing number of women who work in plantation sector but their health status in sadly very poor. They live and work in unhygienic conditions which increase their susceptibility to chronic diseases and infection. Malnutrition coupled with poor health and lack of awareness are critical issues faced by poor women in plantation fields. On account of their poverty they are preoccupied with their daily battles for survival. It is therefore important to bring health care to the centre-stage in the development agenda. Also women should be educated and enlightened to realize their rights. All provisions should be made available by the government to provide adequate health services at affordable cost, so that women in the remote and backward areas can have access to it and thereby lead a healthy life. A healthy woman would make a healthy family and a healthy society.

The innovative manner in which the Chinese went about building their health services in the rural areas is well documented. Their primary emphasis was on ensuring that the basic needs of the population were met, and it is upon this base that preventive campaigns were launched. Health services were developed in a phased manner beginning with primary level care followed by secondary and tertiary levels. However, most writings on China do not give adequate attention to the factors that fell beyond the scope of health services and were far more influential in improving the health status of populations. The reason for citing the Chinese experience is to emphasise the relative role that health services play in contributing to health security and restating the access to basic needs for health improvement (Acharya, Baru and Nambissan, 2001).

The Rise of the Welfare State and Health Services

Until the late nineteenth century period health care was a private and individual responsibility in Europe. The beginnings of state responsibility are seen during the late 1800s when Bismarck introduced a social insurance scheme for workers in Germany. This was then an example that other European countries also followed. State financing and provisioning of health services is a post Second World War phenomenon in Europe that was guided by principles of universality and equity. The world over this period witnessed the rise of the state as key actor in the social sectors. The sixties were the period of growth of welfare states during which state investments grew in several parts of the world. However, the seventies witnessed a world recession and it was the social sectors which suffered from cutback in expenditures.

During the eighties and nineties, there has been an acceleration of privatization throughout the world. This has been due to cutback on public expenditure and expansion of the private sector. These trends were clearly visible in the Indian context when the private sector expanded during the eighties and nineties. In addition to the growth of the private sector, the restructuring of the secondary and tertiary levels of care in the public sector is also a part of the privatization process. If one examines the structure of provisioning at the primary, secondary and tertiary levels of care, then one finds that there is a substantial private sector at the primary level consisting of individual practitioner, both formal and informal. The public sector has sub-centres and primary health centres but its availability and accessibility is variable across states. At the secondary level, the private sector consists of nursing homes with bed strength ranging from 5 to around 50. These institutions are confined to the better-developed states, urban areas and, in some cases, to rural areas. The tertiary sector is confined to the cities, and some of them are corporate enterprises that offer a range of specialist services. Thus, individual practitioners are rendering the bulk of out-patient care in the private sector in in-patient care is mostly by the secondary level institutions. The tertiary sector is being accessed mostly by the upper and middle classes while at the secondary level it is a mixed picture (Baru, 1998).

The high direct and indirect cost of health care often prevents an informal sector worker from seeking treatment, which may result in a worsening of her state of health. The poorest quintile of Indians is 2.6 times more likely than the richest to forego medical treatment when ill, and despite higher rates of illness, is only one-sixth as likely to undergo hospitalization (Peters, et.al, 2001). Poor health, when it results in lost wages and/or health care expenditures, leads to indebtedness, loss of assets and further poverty.

They live and work in unhygienic conditions and are susceptible to many infectious and chronic diseases. A vast majority of them have no fixed employer-employee relationships; nor do they get any statutory social security benefits. Workers in the informal sector do not get health care benefits, nor paid leave for illness, maternity benefits, insurance, old age pension or any other benefits. They receive very low wages, and, as own-account or self-employed workers, they obtain meager piece-rated earnings. At the same time, most workers in the unorganized sector are not members of unions or associations. They thus remain without any representative organizations, which could otherwise help them fight against the many injustices they face everyday. They also do not have the bargaining power or collective strength to demand just policies and laws, including the laws for social protection and social security (Ahmad, et.al., 1991).

Considering the economic conditions of the country, universal coverage or even majority coverage cannot be achieved by either or both the health insurance systems and the government has to continue to play an important role both in terms of financing and delivery, especially for the lower income people and those in the rural areas. However, some resource pressure can be eased from the government by expanding the coverage under both the private and social insurance systems and especially under the latter.

If health security means equitable distribution of health care and low levels of morbidity and mortality, the state has a long way to go. While the health map of Andhra Pradesh, on most indicators, compares poorly with the rest of the southern states, on the reforms front, Andhra Pradesh is in the forefront with significant shifts in health policy. An in-depth assessment of the reforms in the health sector is still early, yet the patterns are discernible. For ensuring health security in Andhra Pradesh, the urgent need is for the state to play a more proactive role in being the major provider of health care services to its people, in both rural and urban areas. Health has to be considered a right and the state has to bear the responsibility of ensuring the well-being of its people.

References

Patnaik, Utsa (2004), "Rural India in Ruins", Frontline, March, 12.

Prasad Sujata (2006), "Health Security: A Fractured Discourse" cited in Securing Health for all. Dimensions and Challenges. (Ed) Sujata Prasad and C. Sathyamala. Institute for Human Development, New Delhi.

Ramachandran, T. (2008), "Rural Women's Health - A study in rural villages of Kanyakumari District", Social Welfare, 55(1-2)

Sandhyavani, B.V., (2008), Health problems of Rural women. Kurukshetra. Oct.

Sathyamala, C (2006), "Health security for the poor: A view from the grass roots", cited in "Securing health for all" op.cit.

Singh, D.P. (2005), "Women workers in unorganized sector", Deep and Deep Publications, New Delhi.

16

Sustainable Development through Horticulture and Fruit Processing

Rupendra R. Gaikwad

Introduction

'The dream is not which one sees in sleep but the dream is that which does not allow to sleep'. The great freedom fighters, who gave their lives for the freedom of their motherland, had a dream of freedom, equality and brotherhood in independent India. India's independence is now more than sixty-one-year-old. What is the extent of achievement of this dream even after these many years? A million dollar question. During last six decades, many governments promised to make the country as the happiest country. If we open the manifestos of all the political parties made for last few elections, we will find many of the issues as repeated. For example, safe and pure drinking water facility, uninterrupted power supply to the micro interiors, road facilities to connect all the villages with urban areas. Overall development of the rural areas is also one of them.

The tragedy is, all these assurances have remained on the paper only. It is not taking a concrete shape. In fact, there is no political will power to make it happen. Therefore, the people from rural areas should think for their own development and without much relying on the government, should make their own ways. Majority of the villagers from rural areas have agriculture as their main activity. Agriculture has got its own difficulties and limitations. As our agriculture activities wholly depend on the monsoon, the results and hence income is not assured. This is the reason why farmers have to go to the moneylenders to meet their immediate needs. When moneylenders come in the picture, the farmer becomes a victim of this vicious cycle of private loans and at the end, suicide.

All these problems are associated with agriculture. Therefore, it is now important to look into this matter seriously. The experts like Dr. Swaminathan and Union Minister for Agriculture, Mr. Sharad Pawar always keep on saying that dependability on the agriculture should be minimized from 100% to 50%-60%, and the focus should be diverted to horticulture. It was rightly said that agriculture was the first green revolution. Now horticulture should be the next. We should go beyond this and must think of processing of horticultural produce. If we can process these horticultural produce on the site itself, i.e., at the point of production, then we can produce quality products and will definitely generate the employment for local people.

Hilly areas are full of natural resources. Only thing needed is proper usage of it with innovative planning. This demands vision. But due to lack of vision and proper planning, no development can take place. Political will power is very much required. A thought was also given to develop tourism in the hilly areas but it requires development of infrastructural facilities, which is again a challenge. We all know the challenges faced by the Konkan Railways.

Horticulture is an activity which can be done with available resources in the hilly areas. This needs a series of episodes of good training programmes. These training programmes can be focused on awareness first of all and then the advantages of it. One more thing will have to be convinced to the farmers of hilly areas that horticulture is possible along with agriculture. This means that horticulture is not a substitute to agriculture but it is a co activity. Normally, hilly regions get heavy rainfalls. Still due to lack of proper water harvesting facilities, they face water scarcity during summer season. If water resource is managed in an innovative manner, it can be utilized to develop horticultural activity.

Processing of horticultural produce is a next step for further prosperity. The tendency of farmers regarding horticulture is selling fresh fruits in the market. As long as good rates are offered in the market, they continue selling fresh fruits. When rates are dropped in the market, they go for processing. This situation is an example of the Konkan region.

The total production of fruits in the world is around 370 million MT. India ranks first in the world with an annual output of 32 million MT. While there are almost 180 families of fruits that are grown all over the world, citrus fruits constitute around 20% of world's total fruit production. The fruits are processed into various products such as fruit juices & concentrates, canned fruit, dehydrated fruit, jams & jellies etc. the levels of processing in the major fruit producing countries in the world are,

- Malaysia : 83%
- Brazil : 70%
- USA : 60-70%
- Israel : 50%

The international trade in processed fruit products is around US$ 9200 million.

India with its current production of around 32 million MT accounts for about 8% of the world's fruit production. Due to the diverse agroclimatic zones, the country makes it possible to grow almost all varieties of fruits and vegetables in India.

Although India is the largest producer of fruits in the world, the production per capita is only about 100 gms per day. However, it is estimated that more than 20-22% of the total production of fruits is lost due to spoilage at various post-harvest stages. Thus the per capita availability of fruits is further reduced to around 80 gms per day which is almost half the requirement for a balanced diet.

The fruit production in India has recorded a growth rate of 3.9%, whereas the fruit-processing sector has grown at about 20% per annum. However, the growth rates have been extensively higher for frozen fruits & vegetables (121%) and dehydrated fruits & vegetables (24%). There exist over 4000 fruit processing units in India with an aggregate capacity of more than 12 lakh MT (less than 4% of the total fruits produced). It is estimated that around 20% of the production of processed fruits is meant for exports, the rest caters to the defence, institutional sectors and household consumption. Mango and mango-based products constitute 50% of exports.

All over the world, processing is done extensively. Brazil is one of the developing countries like India. But there the processing rate is about 70 per cent, whereas in India it is hardly 1.5 to 2 per cent. Many other countries also process about 60-70 per cent fruit production. The business of food processing has vast potential for growth. If we could develop food-processing industry, it will generate enormous employment opportunities. Processing, sales, distribution, transportation will attract the manpower. This will really bring revolution in India and it will make agriculture as a profitable business with the touch of commercialization.

Growing Demand for Processed Food

As an impact of urbanization, lifestyle of middle class has been changed. It is a need of time that both the partners in a family are working. Ladies are out of their house for more than 8 hours daily. In urban areas, life has become so fast that people get less time at their homes. Whatever time they get at home, they want to live life king-size. Therefore, cooking at home has become unwanted activity. As a result, the demand for readymade food or ready-to-cook food is increasing. The media is also playing a very important role in creating awareness as well as attraction towards processed food. This shows that there is a vast market available domestically. In more than 100 crores of population, more than 25 per cent is a middle class, which is becoming potential in near future. This population is bigger than Europe and America. Young generation of Indian farmers should not overlook this much big market. This potentiality has been smelled by MNCs. Therefore, some MNCs have already started tapping Indian market. E.g. Nestle, HUL, etc. Farmers need to grab this opportunity for self-employment and entrepreneurship.

The market potential is unlimited for exports. Statistics show that during 1991, we exported processed food stuffs of worth Rs. 102 crores. That had been marginally increased to Rs. 1366 crores during 2001. This statistic is enough to underline the growing importance of food processing.

Small Saving Groups

As the demand for processed food is increasing, it becomes our responsibility to fulfill this demand. This task can be achieved with the help of small saving groups or on cooperative basis. Small cooperative societies can be formed processing units can be formed under their supervision. The society should shoulder the responsibility of training and convincing of local growers. Knowledge of marketing to the society is important. Here government can help the societies for marketing of the finished goods. If MSAMB help the farmers, middlemen can be eliminated and the benefit can be passed on to the farmers as well as customers. As these products are produced by cooperative societies, government should give some rebate in the taxes so as to keep the prices lower than other competitors. Facility of cold storages should be provided by the government. If all these things are made available, the fruits like Kokum, Jackfruit, Karwanda, Jamun can be processed and prosperity to the hilly regions can be expected.

References

Sen Amithabha, Food Processing Industry in India, India's food processing industry: An IndiaOneStop.com report, July 11, 2005.

Kharche R.M., Food Processing Industry in Maharashtra, Monthly Economic Digest, August 2005.

Kulkarni D.N., Food Processing Industry: A sunrise industry, Shetkari, November 2005.

Magar Shankarrao, Agri Produce processing and exports, Shetkari, Jan-Feb 2002.

Vilaschandran T., Food Processing Industry in Maharashtra, July, 16, 2005, Fruits & Vegetables: Technology Status & Future Vision.

Phalbag (Marathi), Phalbag Vrutta, March-April 2005.

Krushi Panan Mitra (Marathi), MSAMB, March 2005.

17

Sustainable Agricultural Development in India: A Perspective

A. Marimuthu & K. Meenatchi Sundaram

Introduction

India has successfully produced 191 million tones of food grains during 1997-98. With the present growth rate is population and income, the total food grains requirement of estimated at 220.5 and 243.2 million tones in the end of Ninth and Tenth plan. This could be possible through intensive cultivation practices including introduction of high yielding varieties and use of chemical fertilizers, Pesticides and fungicides.

Agriculture accounts for 22 per cent of the GDP and provides direct employment to 58% of the country's population. Indian agriculture is fully dependent on monsoons. Agricultural sector in India is reeling under the pressure of natural calamities on the one hand and starving for credit and modernization on the other.

The Agricultural sector registered a rate of growth at 9.07% in 2003-04 as against 5.2% in 2002-03. The average agricultural GDP growth in the first two years of the plan was 1.8% and it its unlikely exceed 1.5 per cent an the current year. Therefore, the fundamental challenges before us in the 21st century are to find ways and means for sustainable development that are environmentally sound, equitable and allow respect for individual and social rights.

Meaning of Sustainable Development

Sustainable development is a process in which development can be sustained for generations. It is a development which affords to the future generations the same, if not more, capacity to prosper as the present generation has.

In this well written report "Our Common Future", the concept of sustainable development is development that meets the needs of the present without compromising the ability of future generations to meet their own needs.

Contribution of Indian Agriculture

India, which is one of the largest agriculture based economics, remained closed until the early 1990s, what Indian agriculture need is a "Level Playing Field" with domestic industry, in an open economy environment. The Approach paper to the Tenth Fiive year plan advocated that India should accelerate its domestic reforms to create conditions for competitive advantage by domestic and foreign invested enterprises. However, what seems to be a major shift form India's long standing objective of self-reliance in food grains, as the Approach paper suggests.

Of late, in the discussion regarding sectoral policies for the eleventh plan a drastic change is envisaged, specifically for accelerating growth in agriculture to crowd 4 per cent. The demand side intervention such as the recently introduced National rural Employment Guarantee Programme, expanding access and improving quality of Public sector schools health facilities, improved rural connectivity envisaged through Bhrat Nirman and strengthening support to agricultural diversification for domestic processing.

On the supply side, no dramatic technological breakthrough comparable to the "Green Revolution" presently insight but the growth in cereals, purses and oilseeds is possible merely through plausible yield increase in currently low yield regions in India for several reason.

Agriculture has to move from stagnation to vibrant growth. There are promising opportunities, conceptualized, articulated and less realized. As to the roles of for supporting economic growth, securing larger income for better livelihood and to absorb unemployed pool of human resources agriculture has to grow. The issues could be sorted out for bringing in substantial changes in the approach and the components of the change are related to

1. Growth and Stability
2. Food Security
3. Regional Equity and Development.

1. Growth and Stability

Agriculture has been considered once as drag on economic growth and development because of its low productivity syndrome. Technological progress through effective R&D system provides support for output growth, and through varietal improvement and choice to get over biotic stress stability is introduced with limited success since monsoons and rains continue to flaunt a game on the farmers.

A high growth performance over a period is constrained due to stochastic weather aberrations which in most of the causes destroy the asset base and sap the process of cumulative growth in investment. Monsoon failure for year cause not only the loss of current income but also liquidation of productive assets created over a decade or so. Irrigation is one of the productive infrastructure and the potential if 110 million hectares, later upgraded to 143 million hectares taking into consideration new technologies and management would introduce stability in agricultural productivity and output.

2. Food Security

The role of the state in most of the populous and developing countries in providing food security to the teeming million can not be fully market oriented, nor the policy instruments shaped for procurement, storage and buffer market operation can be said market-friendly. Currently, procurement of wheat and rice has been quiet successful and the demand of the Public Distribution System could be met. The basic question of the cost of effective storage and dispersal of stocks of foods grains and the probable cost of alternate system needs to be assessed, analyzed and compared for cost effectiveness for a viable role for the system. A number of issues such as food subsidy to keep the consumer and politicians at ease, contracting for imports with considerable lag in time, adequacy of storage and distribution network would need to be tackled both at political, and administrative and organizational levels. We may require new approach for managerial solutions, as of warehousing and transport models for inventorying and distribution.

3. Regional Equity and Growth

Due to differential in resource endowment agricultural development varies widely the regions of India. Series of studies on agro-climatic regional planning clearly indicate wide variations because of irrigation and types of production systems and institutions. The triangle of the Central India covering part of UP, parts of Bihar and Orissa, and MP have tremendous potential and stands to be exploited through development of land and water resources. Investments and institutions play key role ion its transformation. With regional focus and emphasis on investment for development of land ad water resources it would be possible to reduce the regional imbalances band bring forward the regions on neglect to the new life of growth and development.

Concept of Sustainability

Sustainability of agriculture has been interpreted in an number of ways in the recent literature. It is however, not a new concept. To ensure sustainability in Indian agriculture major changes are needed in two major fields. They are land and soil management and water resource management. India's land resources are under immense pressure as reflected in the fact that it shares only 2 per cent of the world geographical area. But supports about 18 per cent of world population and about 15 per cent of world livestock. The over straining of land resources and disproportionate human and animal population is at the root of widespread land degradation and non-sustainability of an Eco system.

Shift from the agriculture practices to sustainable agrarian system will need a number of measure. Important among such measures are listed here:

1. Continuous soil condition monitoring should be undertaken for assessing their suitability for particular types of crops.
2. Wasteland management should be undertaken on more scientific and ecologically sound manners os that the capacity of such areas maybe enhanced.
3. Soil fertility is the basic capital for all types of agriculture pursuits. So, it should be maintained and protected through cost effective methods.
4. The biological structure of the soil is in danger due to regular us of chemicals, fertilizers and pesticides. It should be used only in limited amount with necessary precautions because it changes physical, chemical and biological composition of the soil.

Water Resource Management

Like the landscape and soil management, scientific and ecologically sound water resource management system has to be evolved for the successful sustainable agrarian development. A properly framed, scientifically managed and conservation oriented water resource management system should include:

1. Water management for the agriculture sector should be based not only on the requirements of the crops but also must take biological properties of the soils to check the soil degradation and erosion.
2. Water resource management should be organized on the watershed and command area basis because it will give an assessment of the availability

and requirement of water and manage cultivation according to this situation.

3. Pollution of underground water resources should be protected. The pollution of ground water is becoming one of the most dangerous problem.
4. Water resource management for the agrarian sector should also take recent scientific development in water utilization in limited quantity and thus conserve water resource. Among many such methods sprinkler and drip irrigation systems are more important.

Policies for Sustainable Agriculture Development in India

Land Use

Sustainable land use refer to the use of land in such away that it meets the needs of both present and future generation which implies that the present generation does not benefit at the cost of future generation. Certain type of land use may appear to be highly profitable in the short run but leads to serious land degradation in the long run, cost of such consequences would outweigh the short run profitability. Hence land use planning plays an important role in optimizing land use to achieve the goal sustainability. Priority should be given for soil conservation hitherto concentrated only on technical aspects and engineering structures with very little attention paid to farmers attitudes and perceptions leading to non involvement of farmers in those programme, Investment analysis clearly indicated the economic viability of soil conservation only at longer time horizons. But the concern of the vast majority of the small and marginal formers is short term profitability of investments though conservation of soil and sustainability of production are of vital importance for social well being.

Water Use

Water scarcity is one of the important constraints in sustaining yield growth. The solution lies in conservation of water and improving the efficiency of irrigation. The existing policy of subsidizing canal water and electricity for irrigation does not encourage facdtorises to conserve water and improving the efficiency of water use in agriculture. Various studies indicated the of improving water efficiency. A study on impact of electricity tariff an ground water irrigation revealed that pricing electricity at 0.50 rupee/KWH would result in a saving of 2 per cent of electricity and ground water use and improvement in water use efficiency about 32 per cent with out any significant reduction in farm income. Hence suitable electricity pricing policy is required to save ground water and electricity Constraints in adoption of drip irrigation and other water saving technologies have to be removed by suitable policy measures.

Energy Use

Indian agriculture is going through a period of rapid change with greater use of commercial forms of energy and a reduction in the share of muscle energy. The growing use of chemical fertilizers, pesticides and groundwater drew through growing use of commercial every needs to considered fully, particularly if future strategies have to be developed to create sustainable patte4n development. Appropriate energy pricing policy along with technological intervention may encourage farmers adopt energy saving strategies leading to improvement in farm machineries such as tractors and pump sets will help in conserving diesel and electricity. Fertilizers and pesticides, the energy intensive inputs play an important role in augmenting crop yield.

Conclusion

In sum, the major issues of agricultural development in the first two decades of the next millennium re concerned with resource development, conservation and use in sustainable with inter-generational equity in focus. The issues of growth and development in agriculture, food security, regional development, continue to be the challenges with the changed focus. Participatory development with the assistance of Panchayats institutions, and Non-Government Organization would motivate greater involvement of people who matter institutions old and new, need to be fostered and developed.

References

Kanchan Chopra (1993), "Sustainability of Agriculture", *Indian Journal of Agricultural Economics*, Vol. 48, No. 3, pp. 516-526.

Nadkarni, M.V.(1993), "Sustainability Development: Concept and issues with special reference to agriculture", *Indian Journal of Agricultural Economics*, Vol. 48, No. 3, pp. 516-526.

Singh, Katar (1998), "Sustainable Development: Some Reflections," *Indian Journal of Agricultural Economics*, Vol. 54, No.1, pp.6-43.

Sudhir Sharma & Babita (2007), "Contribution of Indian Agriculture in India's Trade in Post Reform Period," *Southern Economist*, Vol. 45, No. 19, pp. 5-8.

Valsamma & Antony (2007), "Towards sustainable Development of the Agriculture Sector", *Southern Economist*, Vol. 46, No. 10, pp. 32-34.

18

Sustainable Agricultural Development: Some Issues

Jagannath, B. Kukkudi

Sustainable agriculture are generally concerned with the need for agricultural practices to be economically viable, to meet human needs for food, to be environmentally positive, and to be concerned with quality of life. Since these objectives can be achieved in a number of different ways, sustainable agriculture is not linked to any particular technological practice. Nor is sustainable agriculture the exclusive domain of organic farming. Rather, sustainable agriculture is thought of in terms of its adaptability and flexibility over time to respond to the demands for food and fiber (both high and low), its demands on natural resources for production, and its ability to protect the soil and the resources. This goal requires an efficient use of technology in a manner conducive to sustainability. Finally, because agriculture is affected by changes in market and resource decisions in other sectors and regions, it is important that these changes do not provide a rationale for depleting the agricultural resource base locally.

'Sustainability' is one of the buzz-words of our times. It has been attached to any number of social and economic pursuits: we want sustainable economic growth, sustainable development, sustainable forestry, a sustainable population, sustainable cities, and so on. Increasingly, we also talk about the need for "sustainable agriculture". But just because we use the word in abundance doesn't mean we're any closer to achieving it in practice, or even in understanding what it means. In agriculture, we need to define it in terms that farmers can understand and to develop ways in which they can measure it. Above all, perhaps, we need to ensure that the concept is useful to farmers.

What is Sustainable Agriculture?

The idea of sustainable agriculture has been around a long time. Since the very first crop was sown and animal was penned, farmers have tried to ensure that their land produces a similar or increasing yield of products year after back-breaking year; recent attempts to popularize the concept build on this tradition. One modern definition is as follows: Sustainable agriculture is the use of farming systems and practices which maintain or enhance:

1. the economic viability of agricultural production;
2. the natural resource base; and
3. Other ecosystems which are influenced by agricultural activities.

This definition can be supplemented by some fundamental principles of sustainable agriculture:

1. that farm productivity is enhanced over the long term;
2. that adverse impacts on the natural resource base and associated ecosystems are ameliorated, minimized or avoided;
3. that residues resulting from the use of chemicals in agriculture are minimized;
4. that net social benefit (in both monetary and non-monetary terms) from agriculture is maximized; and
5. That farming systems are sufficiently flexible to manage risks associated with the vagaries of climate and markets.

Developing a List of Indicators

The fact is, sustainability cannot be measured directly; it is too elusive a concept and it operates over too long a time scale. The best we can do is to identify measurable phenomena that, when put together, suggest how sustainable our system might be. These are called indicators. Indicators are widely used as benchmarks to help gauge performance in a number of human endeavours. For example, the consumer price index and gross domestic product are indicators, albeit crude ones, of economic performance.

In Australia, some work has already been done on the development of indicators for sustainable agriculture. In 1992, the Standing Committee on Agriculture and Resource Management (SCARM) established an expert group to develop a set of indicators to be used by decision-makers at the regional and national scales. This led to the establishment of a National Collaborative Project on Indicators of Sustainable Agriculture, which aims to prepare 'report cards' on the sustainability of Australian agriculture. A number of criteria can be used to judge the usefulness of a given indicator:

1. Is it measurable?
2. Is it relevant and easy to use?
3. Does it provide a representative picture?
4. Is it easy to interpret and does it show trends over time?
5. Is it responsive to changes?
6. Does it have a reference to compare it against so that users are able to assess the significance of its values?
7. Can it be measured at a reasonable cost, and can it be updated?

What Farmers Think of Indicators?

The coordinators of the sustainability indicators project analyzed the SCARM indicators and constructed user-friendly interpretations of them. They also compiled a list of possible on-farm indicators from a range of sources and produced a draft guide that discussed and explained their use. The draft guide was then used in discussion groups with FM500 members as a basis for debating sustainability policies in general, and the various indicators in particular. Comments on the usefulness of each indicator were received from about 180 farm businesses. These results suggest that FM500 farmers generally agreed with the potential value of all indicators at both scales. The lowest ranked national/ regional indicator, 'length of contact with non-agricultural areas', achieved an average rating of just under 3, implying that even this indicator was considered useful, if not of major importance.

Farmer groups sometimes varied in their views, depending on the type of enterprises represented or the environments in which they operated. For example, farmers from regions with typically alkaline soils generally rated per centage organic carbon in the soil more highly than they did soil acidification. The interests of the facilitator or the nature of the agricultural enterprise could explain other variations. For instance, groups facilitated by a consultant whose focus was on cropping tended to value indicators such as water use efficiency (WUE) more highly than groups from grazing areas where the WUE concept was not so well developed. According to the project coordinators, the general agreement of FM500 farmers on the relevance of indicators was not surprising, given that they were intimately involved with their development. In addition, the high rating given to profitability as a measure of sustainability could reflect the bias of FM500 facilitators - but may also reflect the fact that such indicators are generally better developed and understood than others. Indeed, the high ratings given to almost all indicators are they orientated towards financial, environmental or management issues suggest that farmers are not averse to taking a holistic view of farm performance.

Making Indicators Relevant to Farmers

The process of consulting with farmers about sustainability indicators was just as informative as the outcomes themselves. Early on, facilitators found it necessary to 'bring the indicators to life' by explaining their background and justifying the need for them. Without such justification, say the project coordinators, farmers were profoundly uninterested! So, why should farmers pay any attention to indicators of sustainability? The project coordinators advanced five possible reasons:

1. indicators can help farmers notice changes at an early stage and seek advice if required;
2. profitability indicators can highlight strengths and weaknesses and show trends;
3. land and water quality indicators can highlight natural resource issues which may be 'sleepers' and not obvious to the eye until they are well advanced and difficult to address;
4. managerial skills self-auditing can assist individual business partners to appraise honestly their talents and to plan for professional development; and
5. Off-site impact monitoring can ensure that individual businesses maintain quality standards and do not contribute to problems for the wider community.

Indicators can play an important role in regional policymaking and in helping Australia meet its international obligations. They may also prove useful to other groups. For example, the initial response of agribusinesses and educational organizations to the development of a sustainability indicators guide was extremely positive. They saw it as a tool with which to educate staff in sustainable resource management issues.

The End Result

The main product of all this work is a recently published booklet: *Sustainability indicators for agriculture*. It provides information on indicators at two different scales: that relevant to regional/national rural industry policymaking, and that relevant to on-farm management decisions. It is intended as a learning tool by which farmers and others can familiarize themselves with issues relevant to sustainable agriculture.

The booklet is certainly not intended as the last word on sustainable agriculture: it will feed into a wider process of indicator development which itself is only part of the quest for sustainability. But indicators help us to define that buzz-word, *sustainability*, and to work out what more we need to do to achieve it.

If advocating the need for a sustainable agriculture has become universal, agreement as to what is required to achieve it has not. As more parties sign on to the sustainable agriculture effort, perceptions about what defines sustainability in agriculture have multiplied. This paper strives to illustrate the commonality and some of the controversy that defining such a goal entails, and it includes brief descriptions of the methodologies and practices currently associated with sustainable agriculture.

Sustainable Agriculture: The Basics

Some terms defy definition. "Sustainable agriculture" has become one of them. In such a quickly changing world, can anything be sustainable? What do we want to sustain? How can we implement such a nebulous goal? Is it too late? With the contradictions and questions has come a hard look at our present food production system and thoughtful evaluations of its future. If nothing else, the term "sustainable agriculture" has provided "talking points," a sense of direction, and an urgency, that has sparked much excitement and innovative thinking in the agricultural world.

The word "sustain," from the Latin sustinere (sus-, from below and tenere, to hold), to keep in existence or maintain, implies long-term support or permanence. As it pertains to agriculture, sustainable describes farming systems that are "capable of maintaining their productivity and usefulness to society indefinitely. Such systems... must be resource-conserving, socially supportive, commercially competitive, and environmentally sound. Under the law, "the term sustainable agriculture means an integrated system of plant and animal production practices having a site-specific application that will, over the long term:

1. satisfy human food and fiber needs;
2. enhance environmental quality and the natural resource base upon which the agricultural economy depends;
3. make the most efficient use of nonrenewable resources and on-farm resources and integrate, where appropriate, natural biological cycles and controls;
4. sustain the economic viability of farm operations; and
5. Enhance the quality of life for farmers and society as a whole."

Some Background

How have we come to reconsider our food and fiber production in terms of sustainability? What are the ecological, economic, social and philosophical issues that sustainable agriculture addresses? The long-term viability of our current food production system is being questioned for many reasons. The news media regularly present us with the paradox of starvation amidst plenty—including pictures of hungry children juxtaposed with supermarket ads. Possible adverse environmental impacts of agriculture and increased incidence of foodborne illness also demand our attention. "Farm crises" seem to recur with regularity.

The prevailing agricultural system, variously called "conventional farming," "modern agriculture," or "industrial farming" has delivered tremendous gains in productivity and efficiency. Food production worldwide has risen in the past 50 years; the World Bank estimates that between 70 per cent and 90 per cent of the recent increases in food productions are the results of conventional agriculture rather than greater acreage under cultivation. U.S. consumers have come to expect abundant and inexpensive food.

Philosophical underpinnings of industrial agriculture include assumptions that "(1) nature is a competitor to be overcome; (2) progress requires unending evolution of larger farms and depopulation of farm communities; (3) progress is measured primarily by increased material consumption; (4) efficiency is measured by looking at the bottom line; and (5) science is an unbiased enterprise driven by natural forces to produce social good.

Significant negative consequences have come with the bounty associated with industrial farming. Concerns about contemporary agriculture are presented below. They are drawn from the resources compiled at the end of this chapter. While considering these concerns, keep the following in mind: (1) interactions between farming systems and soil, water, biota, and atmosphere are complex—we have much to learn about their dynamics and long term impacts; (2) most environmental problems are intertwined with economic, social, and political forces external to agriculture; (3) some problems are global in scope while others are experienced only locally; (4) many of these problems are being addressed through conventional, as well as alternative, agricultural channels; (5) the list is not complete; and (6) no order of importance is intended.

Ecological Concerns (Negative effects of current practices)

1. Decline in soil productivity can be due to wind and water erosion of exposed topsoil; soil compaction; loss of soil organic matter, water holding capacity, and biological activity; and salinization of soils and irrigation

water in irrigated farming areas. Desertification due to overgrazing is a growing problem, especially in parts of Africa.

2. Agriculture is the largest single non-point source of water pollutants including sediments, salts, fertilizers (nitrates and phosphorus), pesticides, and manures. Pesticides from every chemical class have been detected in groundwater and are commonly found in groundwater beneath agricultural areas; they are widespread in the nation's surface waters. Eutrophication and "dead zones" due to nutrient runoff affect many rivers, lakes, and oceans. Reduced water quality impacts agricultural production, drinking water supplies, and fishery production.
3. Water scarcity in many places is due to overuse of surface and ground water for irrigation with little concern for the natural cycle that maintains stable water availability.
4. Other environmental ills include over 400 insects and mite pests and more than 70 fungal pathogens that have become resistant to one or more pesticides; stresses on pollinator and other beneficial species through pesticide use; loss of wetlands and wildlife habitat; and reduced genetic diversity due to reliance on genetic uniformity in most crops and livestock breeds.
5. Agriculture's link to global climate change is just beginning to be appreciated. Destruction of tropical forests and other native vegetation for agricultural production has a role in elevated levels of carbon dioxide and other greenhouse gases. Recent studies have found that soils may be sources or sinks for greenhouse gases.

Economic and Social Concerns

Economic and social problems associated with agriculture can not be separated from external economic and social pressures. As barriers to a sustainable and equitable food supply system, however, the problems may be described in the following way:

1. Economically, the U.S. agricultural sector includes a history of increasingly large federal expenditures and corresponding government involvement in planting and investment decisions; widening disparity among farmer incomes; and escalating concentration of agribusiness—industries involved with manufacture, processing, and distribution of farm products—into fewer and fewer hands. Market competition is limited. Farmers have little control over farm prices, and they continue to receive a smaller and smaller portion of consumer dollars spent on agricultural products.

2. Economic pressures have led to a tremendous loss of farms, particularly small farms, and farmers during the past few decades—more than 155,000 farms were lost from 1987 to 1997. This contributes to the disintegration of rural communities and localized marketing systems. Economically, it is very difficult for potential farmers to enter the business today. Productive farmland also has been pressured by urban and suburban sprawl—since 1970, over 30 million acres have been lost to development.

Impact on Human Health

Potential health hazards are tied to sub-therapeutic use of antibiotics in animal production, and pesticide and nitrate contamination of water and food. Farm workers are poisoned in fields, toxic residues are found on foods, and certain human and animal diseases have developed resistance to currently used antibiotics.

Philosophical Considerations

Historically, farming played an important role in our development and identity as a nation. From strongly agrarian roots, we have evolved into a culture with few farmers. Less than two per cent of Americans now produce food for all U.S. citizens. Can sustainable and equitable food production be established when most consumers have so little connection to the natural processes that produce their food? What intrinsically American values have changed and will change with the decline of rural life and farmland ownership? World population continues to grow. According to recent United Nations population projections, the world population will grow from 5.7 billion in 1995 to 9.4 billion in 2050, 10.4 billion in 2100, and 10.8 billion by 2150, and will stabilize at slightly under 11 billion around 2200. The rate of population increase is especially high in many developing countries. In these countries, the population factor, combined with rapid industrialization, poverty, political instability, and large food imports and debt burden, make long-term food security especially urgent.

Finally, the challenge of defining and dealing with the problems associated with today's food production system is inherently laden with controversy and emotion. "It is unfortunate, but true, that many in the agriculture community view sustainable agriculture as a personal criticism, or an attack, on conventional agriculture of which they are justifiably proud. I guess that the main thing people get defensive about when you say sustainable,' explained one agent, 'is that it implies that what they've been doing is not sustainable. And that's the biggest issue.'

A Sampling of Perspectives

"Sustainable agriculture is a philosophy based on human goals and on understanding the long-term impact of our activities on the environment and on other species. Use of this philosophy guides our application of prior experience and the latest scientific advances to create integrated, resource-conserving, equitable farming systems. These systems reduce environmental degradation, maintain agricultural productivity, promote economic viability in both the short and long term, and maintain stable rural communities and quality of life. Sustainable agriculture does not mean a return to either the low yields or poor farmers that characterized the 19th century. Rather, sustainability builds on current agricultural achievements, adopting a sophisticated approach that can maintain high yields and farm profits without undermining the resources on which agriculture depends.

"A systems approach is essential to understanding sustainability. The system is envisioned in its broadest sense, from the individual farm, to the local ecosystem, and to communities affected by this farming system both locally and globally... A systems approach gives us the tools to explore the interconnections between farming and other aspects of our environment. Environmental sustainability implies the following:

1. Meeting the basic needs of all peoples, and giving this priority over meeting the greeds of a few.
2. Keeping population densities, if possible, below the carrying capacity of the region.
3. Adjusting consumption patterns and the design and management of systems to permit the renewal of renewable resources.
4. Conserving, recycling, and establishing priorities for the use of nonrenewable resources.
5. Keeping environmental impact below the level required to allow the systems affected to recover and continue to evolve.

Sustainable agriculture is "a way of practicing agriculture which seeks to optimize skills and technology to achieve long-term stability of the agricultural enterprise, environmental protection, and consumer safety. It is achieved through management strategies which help the producer select hybrids and varieties, soil conserving cultural practices, soil fertility programs, and pest management programs. The goal of sustainable agriculture is to minimize adverse impacts to the immediate and off-farm environments while providing a sustained level of production and profit. Sound resource conservation is an integral part of the means to achieve sustainable agriculture."

"Today, sustainable farming practices commonly include:

1. crop rotations that mitigate weeds, disease, insect and other pest problems; provide alternative sources of soil nitrogen; reduce soil erosion; and reduce risk of water contamination by agricultural chemicals
2. pest control strategies that are not harmful to natural systems, farmers, their neighbors, or consumers. This includes integrated pest management techniques that reduce the need for pesticides by practices such as scouting, use of resistant cultivars, timing of planting, and biological pest controls
3. increased mechanical/biological weed control; more soil and water conservation practices; and strategic use of animal and green manures
4. use of natural or synthetic inputs in a way that poses no significant hazard to man, animals, or the environment.
5. "Sustainable agriculture is a model of social and economic organization based on an equitable and participatory vision of development which recognizes the environment and natural resources as the foundation of economic activity. Agriculture is sustainable when it is ecologically sound, economically viable, socially just, culturally appropriate and based on a holistic scientific approach.
6. "Sustainable agriculture preserves biodiversity, maintains soil fertility and water purity, conserves and improves the chemical, physical and biological qualities of the soil, recycles natural resources and conserves energy. Sustainable agriculture produces diverse forms of high quality foods, fibers and medicines.
7. "Sustainable agriculture uses locally available renewable resources, appropriate and affordable technologies and minimizes the use of external and purchased inputs, thereby increasing local independence and self sufficiency and insuring a source of stable income for peasants, family and small farmers and rural communities. This allows more people to stay on the land, strengthens rural communities and integrates humans with their environment.
8. "Sustainable agriculture respects the ecological principles of diversity and interdependence and uses the insights of modern science to improve rather than displace the traditional wisdom accumulated over centuries by innumerable farmers around the world.

The Future of the Sustainable Agriculture Concept

Many in the agricultural community have adopted the sense of urgency and direction pointed to by the sustainable agriculture concept. Lack of sharp definition has not lessened its authenticity. Sustainability has become an integral component of many government, commercial, and non-profit agriculture research efforts, and it is beginning to be woven into agricultural policy. Increasing numbers of farmers and ranchers have embarked on their own paths to sustainability, incorporating integrated and innovative approaches into their own enterprises. This just-do-it attitude is the real force carrying the issue of sustainability into the next century. "The best way to communicate the meaning of sustainable agriculture is through real-life stories of farmers who are developing sustainable farming systems on their own farms. Critical discussion of the sustainable agriculture concept will and should continue. Understanding will deepen; answers will continue to come. On-going dialog is important for another reason: with more parties, each with its own agenda, jumping into the sustainable agriculture "tent," only a continued focus on the real issues and goals will keep sustainable agriculture from becoming so all-encompassing as to become meaningless.

References

Subchapter I: Findings, Purposes, and Definitions, *U.S. Code, Title 7, Chapter 64-Agricultural Research, Extension and Teaching, Available at GPO Access:*

Charles Francis and Garth Youngberg, "Sustainable Agriculture — An Overview," in *Sustainable Agriculture in Temperate Zones,* edited by C.A. Francis, C.B. Flora and L.D. King (New York: Wiley, 1990

Karl N. Stauber *et al.,* "The Promise of Sustainable Agriculture," in *Planting the Future: Developing an Agriculture that Sustains Land and Community,* Elizabeth Ann R. Bird, Gordon L. Bultena, and John C. Gardner, editors (Ames: Iowa State University Press, 1995).

19

Scenario of Horticulture Development in Karnataka

Gouradevi Katnalli

Introduction

In the era of liberalization, privatization and globalization, it is essential to attain the sustained or long lasting development of the economy. Agriculture being the backbone of Karnataka economy it is still more important to attain and maintain the sustained development of Agriculture, which meets not only the needs of present generation but also the needs of future generation.

The sustainable development of agriculture largely depends upon the development of allied sectors like plantation, horticulture, sericulture etc. In this paper an attempt is made to analyze the scenario of the development of horticulture in Karnataka.

Objectives

The main objectives of the study are

1. To review the present position of the development of horticulture in Karnataka.
2. To study the major strengths and weakness of the state in relation to horticultural development.
3. To analyze the major steps taken by the government for the development of horticultural sectors.
4. To study the strategy for the development of post-harvest infrastructure and management.
5. To study the future thrust of the state.

Role of Horticulture in Karnataka

Karnataka occupies a prominent place in the horticulture map of the country. The horticulture crops are grown in an area of 15.80 lakh ha with a total production of over 124.00 lakh tones at the end of VIII plan period. At present its estimated production is 95.81 lakh tones at the end of 2002-03. Although the area comprises only 14.44% of the net cultivated area in the state, the total income generated from the horticulture sector accounts for over 40% of the total income derived from the combined agriculture sector. This accounts for 17% of the GDP of the state.

Horticulture provides excellent opportunities in raising the income of the farmers even in the dry tracts. A significant shift towards horticulture is evident in the state with an increase in area and production. For instance, about 58,000 ha area has been brought under horticultural crops through the watershed programmes. Horticulture provides higher unit productivity and offers great scope for value addition. Karnataka having the highest acreage under dry farming in the country, next only to Rajasthan, has a great potential to grow high-value but less water-demanding horticultural crops.

Major Horticultural Crops

Karnataka is a progressive state in the field of modern horticulture in the country. The diverse agro-ecological conditions prevailing in Karnataka has made it possible to grow different types of horticultural crops such as fruits, vegetables, flowers, species, plantation crops, root and rubber crops, medicinal and aromatic crops, etc. Karnataka is the 3rd-largest producer of fruits and stands in the 5th position in area and production of vegetable crops. The state stands 1st in area and 3rd in production of flower crops. It occupies 2nd and 3rd position with respect to area and production of plantation crops respectively. Karnataka is the largest producer of spices, and aromatic and medicinal crops.

For the improvement of any sector in the economy needs a proper planning, knowing and understanding our strengths and weakness, efforts are to be made for the proper and fuller utilization of our strengths and over coming our weakness and capturing and encashing the opportunities.

Strengths

The major strengths of the state of Karnataka in the field of horticulture are as follows:

1. Comparative advantage with respect to the diverse agro-climate condition of the state suitable for growing different types of horticultural crops.

2. Comparative advantage of the geographical location of the state ensures good prospects for marketing (export) of horticultural produce to other states of the country and abroad.
3. Availability of abundant waste and marginal land that can be utilized for cultivation of dry land horticultural crops.
4. More enterprising farmers.
5. Economic liberalization policy of the Government has led to the establishment of many multinational companies involved in horticulture sector in Karnataka. This led to the exposure of the farmers to newer varieties of crops resulting in potential marketing opportunities in both domestic and international export of high-yielding crops such as flowers, hybrid vegetables, etc.
6. The state has a wide network of both private and public sectors for supply of seed, fertilizers and other inputs.
7. The setting up of international air cargo services and cold storage at Bangalore airport has resulted in the boosting of the floriculture sector.
8. Presence of upcoming processing and floriculture industries in the state.
9. Strong linkages between agricultural universities and development departments.
10. Availability of abundant skilled/unskilled manpower in an unorganized sector.
11. Partially strong marketing system like HOPCOMS, KHF, KAPPEC, etc.
12. Strong network of Nurserymen Co-operative.

Weaknesses

1. Weak Extension Linkages in the Departments.
2. Inadequate supply of quality planting material.
3. Up gradation of Technical Knowledge.
4. Area expansion on unscientific basis leading to declined productivity and fluctuation in market prices.
5. Inadequate infrastructure for marketing, handling and processing of horticultural produce.
6. Inadequate air cargo and rail cargo facilities in the state.
7. Inadequate facilities for post-harvest handling of horticultural produce.
8. Inadequate power supply and exorbitant power tariff.

9. Entry of multinational companies posing a threat to small and marginal farmers in terms of market competition and prices.
10. Small irrigation resources in the state compared with the neighboring states.
11. Regional imbalance in the development of horticulture within the state.
12. Mono cropping system of cultivation.
13. Increase in prices of plant protection chemicals and fertilizers.
14. Failure in market stability.
15. Complexity in the incidence of pests and diseases.

Current Production Scenario

The Current Production Scenario of Horticultural Crops is presented in the following Table-1.

As per the statistics of 2002-03, the area under horticulture crops was 15.81 lakh ha with an annual production of 95.81 lakh tonnes. Out of this 41% of area and 87% of total production account for fruits and vegetables.

Table 1
Current Production Scenario

	Area (in Lakh ha) Production (in Lakh Tons)				
Crops	**1998-99**	**1999-00**	**2000-01**	**2001-02**	**2002-03**
Fruit Crops					
Area	3.15	3.13	2.61	2.57	2.55
Production	48.27	46.41	41.65	40.29	40.39
Plantation and Spice Crops					
Area	8.64	8.97	8.71	9.26	9.40
Production	12.46	12.25	8.31	8.04	8.94
Vegetable Crops					
Area	2.59	3.29	3.77	3.83	3.67
Production	52.56	43.11	66.54	48.90	44.94
Flower Crops					
Area	0.21	0.21	0.21	0.18	0.18
Production	1.24	1.32	1.57	1.42	1.47
Total Area	17.18	15.6	15.30	15.85	15.81
Total Production	128.07	103.09	118.20	98.76	95.81

Source: Karnataka state department of Horticulture.

In Karnataka only 1% of the total production of fruits and vegetables are processed into different products. About 25-30% of the produce is lost due to improper post-harvest management. To avoid the post-harvest losses, a chain of cold storages, processing and marketing facilities need to be created.

Processing of Fruits and Vegetables

Government of Karnataka is considering fruit and vegetables processing sector as a thrust area for development. With growing urbanization and increasing quality consciousness, the market for processed fruit and vegetable is expected to grow rapidly. Therefore, there is enormous potential for developing agro products.

At present there are 1,533 licensed processing units in the state with annual production of 2.00 lakh tonnes. The processed fruits are mango, grapes, pineapple, papaya, guava and others. The major vegetables.processed are tomato, potatoes, peas, gherkins and others.

Strategies for Development of Post-Harvest Infrastructure and Management

1. Modernization of existing cold storages.
2. The state Government is contemplating to have Horticultural Marketing Complex similar to Kannamangala at Hubli-Dharwad, Mysore and Bijapur.
3. The Government has proposed to establish Agro Parks at Belgaum, Jewargi, Maddur and Chitradurga.
4. Efforts are on to establish Agri Export Zones for crops like grapes, mango, pomegranate, aromatic and medicinal plants and vanilla.
5. The Department is conducting training programme on preservation and processing of fruits and vegetables to ladies and housewives in Bangalore. The Department is planning to extend the same to other urban areas of the State.
6. Training and educating the farmers in the production of quality fruits, acceptable in international markets for export purposes.

The states first ever Agro Food Industrial Policy which revolutionized agriculture, as well as put fresh graded fruits and vegetables into the urban market is expected to generate Rs. 5,000 cores in foreign exchanges over 3 years. This policy is in the offing.

Marketing

The Department of Horticulture acts the facilitator for creation of infrastructure facilities for marketing of fruits and vegetables in the states. The Department of Agricultural Marketing recently established 'Raithra Santhe' wherein the growers/ farmers can bring their fruits and vegetables to the market and sell them directly to the consumers. The Department of Agricultural Marketing also handles the notified fruits and vegetables in the state through APMCs. The other marketing infrastructure in the state are.

1. The Department promoted HOPCOMS.
2. The Department promoted District Horticultural Societies.
3. Karnataka Agro Industries Corporation.

Exports

During the year 2001-02 produces such as gherkins, Bangalore rose onion, cashew nut, coffee, flowers and horticultural processed products were exported to countries like the US, Singapore, Russia, Malaysia, Indonesia, Australia, France, Italy and Japan. The total value of the export is valued at about Rs. 1,517.13 Cores. The details of the horticultural produces exported during 2003-04 from the state are presented in the following table-2.

Table 2

Export of Horticultural Produces during 2003-04

Heads of Exports	(Rs. In Crores)
Coffee and its products	757.57
Cashew and its products	265.72
Processed food	361.25
Species and its products	99.58
Onion	30.84
Fruits	2.17
Total	**1517.13**

Agri Export Zone were created to boost the exports of horticultural produces from the state, to improve the export performance of the state, to provide opportunities to the farmers to earn incremental income, to produce on-farm and off-farm employment generating activities and to earn foreign exchange.

Future Thrust

The Future Thrust of the State in the development of horticulture is as follows:

1. **Zone-specific Cropping Pattern in Horticulture:** Much emphasis has to be laid on achieving higher production through encouragement for zone wise area expansion of horticultural corps, increase in productivity, use of suitable high yielding varieties, use of improved cultural practices and irrigation systems, integrated pest management and improvement of quality of the produce. Unlike field corps, all the horticultural crops cannot be grown in all the zones. Their needs are specific with regard to agro-climatic requirement. Each zone has its own characteristics. Although the horticultural crops are being grown throughout the state covering all the zones, the production and productivity are not up to the potential. Hence, there should be a policy to visualize the zone specific cropping pattern in horticulture.
2. **Strengthening of Domestic Marketing System through Establishment of Modern Horticultural Markets**: Realizing that the efficient terminal markets would stimulate productivity, raise quality standards, reduce losses and ensure consumer access to an increased supply of fresh horticultural produce at reasonable prices, there is a need for establishment of an alternate marketing in the state. The National Diary Development Board with the active support from GOK is setting up a terminal market for fruits and vegetables at Bangalore for horticultural produce. It is proposed to set up 3 or 4 more similar terminal markets in the state. It is also envisaged to prepare a crop specific and location specific marketing plan. The government will encourage and promote contract farming in horticulture corps. Government also proposes to provide minimum support price for horticultural corps in consultation with state agricultural price commission to encourage domestic marketing of horticultural produce.
3. **Promotion of Organic Cultivation of Horticultural Crops for exports:** Organic farming is a silent revolution in the making and is slated to be the next big boom for Karnataka after Information Technology and Biotechnology. The global market for organically grown fruits, vegetables, species, aromatic and medicinal plants has been growing tremendously for the past 5 years. Karnataka with large extent of dry land, where traditional 'natural' farming methods are still being followed, the conversion of these lands into organic farms is much easier. In light of the above, it is very much needed to tap the export potential of organically grown horticulture produces by providing production and market linkages to organic cultivators.
4. **Satellite Mapping of Horticulture Crops with the help of ISRO to Develop a Horticultural Data Bank:** Data base is essential to take up

developmental programmes. The data on area and production of horticulture crops brought out by various departments/ organizations do not match each other. At the time of taking policy decision, the data base plays a crucial role. Therefore, the department proposes to take up one time survey for perennial horticultural crops and also to use the services of ISRO to employ the latest and sophisticated methods to map the area under different horticultural crops in the state. This would help in proper and real estimation of area under horticultural crops in the state. This would further lead to more scientific estimation of yield and production of horticultural crops in the state.

5. **Major Initiative to Conserve Water for Horticultural Crops**: Keeping in view the limited availability of water and depletion of ground water levels, the use of water saving devices needs to be promoted to maximizes overall irrigation potential. The efficient and maximum utilization of available water is possible through micro-irrigation. The State Government intends to bring more horticultural area under micro-irrigation. To meet the demand of farmers for adoption of micro-irrigation system in their orchards, it is proposed to increase the flow of funds towards micro-irrigation through diversion of funds from minor irrigation department 10 percent annually. It is also envisaged to encourage rain water harvesting on farmers field by providing incentives.

6. **Dryland Horticulture:** Drylands in fact are the only hope of the future for feeding the fast increasing population. To improve the economic conditions of the people in the dry tracts and their nutrition and health standards, it is necessary to encourage dryland horticulture through extension support for dryland orchards and support for development of new technologies for dryland horticulture through R & D. The state has given the highest priority for watershed management of farm areas and also tank development.

References

Alagh Y.K. (1999), "Agricultural Policy in the Ninth Plan". Hans E. Karl and B.J. Nanjundappa. Memorial Lectural, IAT, Bangalore.

Bhalla G.S. and Gurmail Singh (2001), "Indian agriculture-For Decades of Development", New Delhi, Sage.

Bhendel M.J. (2000), "Cropping pattern and Resource use Efficiency in Major crops: A Case Study of Karnataka". ADRT Unit, Institute for Social and Economic Change, Bangalore.

R. Sthanumoorthy and P. Sivarajadhanavel (2007), (ed.) "Karnataka Economy – Performance Challenges and Opportunities" The ICFAI University Press.

Karnataka State Department of Horticulture.

20

Towards Sustainable Farming Practices

M.K. Biradar

Introduction

FAO has defined sustainability as "The management and conservation of Natural resource base and the Orientation of Technological and Institutional change in such a manner as to ensure the attainment and combined satisfaction of human needs for present and future generations".

In the Indian Economy agriculture is playing a vital role and hence one can not even think of economic development of India without taking into account the agricultural Sector Sustainable economic growth depends upon the sustainable agriculture.

Sustainable agriculture is a philosophy as well as a system of production. The production system is considered sustainable in log run only if out put produced are lost can be balanced with inputs into the system. The concept of sustainability is a complicated and contested one. For economists it is only a long term sustainability of economic output, income and consumption, for ecologists it is long term Preservation of biosphere.(i e sustainance of people and bio-diversity) and for sociologists and anthropologists sustainable is one that preserves traditional values. In brief sustainable development could be defined as development that meets the needs of present without compromising the ability of the future generation to meet their own needs.

Agriculture is being an important economic activity for various reasons and hence country like India must take into account the need for sustainable agriculture. Sustainable agriculture is yet an another concept that has been debated especially in the context of chemicalisation of agriculture and its harmful impact on biosphere Sustainable agriculture seeks to conserve land, water, plant, and animal genetic resources. It is environment friendly, non-degrading, technically appropriate, economically viable and socially acceptable.

Objective of the Study

The main objective of this paper is to study:

1. Existing conditions and problems of Indian agriculure.
2. Negative impact of modern farming.
3. Importance of shifting from chemical agriculture to organic farming.
4. Merits of natural farming in the view of Food security and ecological balance.
5. Need for sustainable agricultural development.

Methodology

This article is based on secondary data.

Concept of Organic Farming

There are several important factors on the basis of which organic farming has been promoted in the recent years it can improve the quality of food, producing healthy food, which is free from harmful chemicals and likely to be more nutritious. It helps to improve the quality of soil from harm caused by agro-chemicals. It has the potential to significantly reduce farming costs, as cash expenditure on some expensive inputs like chemicals fertilizers, insecticides, herbicides can be avoided.

Organic farming can help to increase the self reliance of farming communities. In many places it is possible for organic farmers to obtain better price for their produce. Organic produce can make available new markets to farmers including export markets.

Organic farming is better for the survival and growth of other forms of life – wild life, farm animals, other domesticated animals including many friendly birds and insects who also contribute to more productive agriculture by promoting pollination, eating harmful insects and in other ways. Organic farming helps to reduce water pollution, occupational health hazards for farmers and farm workers would be significantly reduced.

Though these aspects of organic farming are must be examined against the crucial question of whether it can produce yields comparable to what has been possible by using agri-chemical like fertilizers, insecticides and herbicides. This question is certainly important in the present context of an increasing food crises being reported from many parts of the world. The answer for this question is happily a yes. It is possible for natural agriculture to feed the world.

But to realise this as well as other potentials of natural agriculture it can not be considered in isolation. It is not sufficient to claim that giving up the use of harmful and hazardous chemicals is enough. It is equally important to pay full-attention to the various conditions in which the natural farming flurishes best – soil, water and mosture conservation, cropping systems must conducive to weather and soil conditions, protection and promotion of enough greenary in and around the village, include pasture land, trees, and forests, social harmony and economic equality in the village. It is under these conditions that village communities can best contribute in realising the advantages of organic farming to feed not only the country but also produce surplus for other parts where it is needed.

Any campaign to promote natural farming in India is most welcome but it should take into account this wider reality it is to make a worthy and lasting contribution to the food security, health improvement and protection of environment. The campaign for organic agriculture must be an integral part of wider understanding of programme of healthy, just and sustainable farming and food systems.

Documentation is already available in India of numerous experimental efforts relating to rich experiences and potential of organic farming. The first example of such efforts, Green framing was by the centre for science and environment. The second by ETC consultants, it tittled "LEISA" network for India. A Register of Indian organisation in low external—input and sustainable agriculture.

The third one is a part of larger book on organic farming titled "The organic farming source book. Published by the other India book store Goa, in collaboration with the third world. Net work, Malaysia, Kalpavriksha, (Pune & Delhi) Centre for Indian knowledge systems (Chennai) and Prakriti (Bombay) have also published information on organic farming in India.

Examples of Successful Stories

There are several examples of individual success stories of farmers using low-cost farming practices without any chemicals while maintaining reasonable yields. In fact in some cases the yields are quite high. In others cases yields decline in the short term but pick-up again after some time. In several of these experiments farmers have attempted to promote a harmonious relationship between crops, trees, farm animals, friendly insects, earth warms, micro organisms and birds.

A percept emphasised by most of them is "understanding the way nature works and carry out your work in accordance with the way of nature".

The society for equitable voluntary action (West Bengal) under took an experiment to grow crops without the use of chemicals fertilizers on the fields of 100 farmers in 10 villages. There was a positive response in as many as 85% cases. In SEVA's project area a growing of paddy without using chemicals fertilizers has spread to 2000 hectares.

From the green revolution district of Bijnore (UP) Shor vir singh says after several years of experimenting organic agriculture with a group of farmers. Now I am quite confident, in a years time I can change the land to organic farming without any decrease in production. Anil/Rana/Jan Hith trust won the one world award. This year for promoting organic farming in Meerut (UP).

The save the seed movement in the Garwal hills the traditional seed Bank in Tamil Nadu and the Deccan Development society in (AP) have succeeded in collecting and spreading the use of traditional seeds by an ever increasing number of farmers. In some states, like Maharashtra, Tamil Nadu and else where, networks of organic farms have formed. This will enable more organic farmers to learn from each others experience. Most farmers clearly expressed satisfaction after shifting to organic farming.

In Sept. 1989the US National Academy of sciences published a report which examined 14 farms in the USA that had successfully developed national production methods. This report on Alternative Agriculture said " well managed alternative farms use less synthetic, chemical fertilizers, pesticides and antibiotics, without necessarily decreasing and in some cases increasing per acre crop yields and productivity of live stock systems without adoption of proven alternative systems would result in ever greater economic benefits to farmers, and environmental gains for the nation.

According to the case study of successful vigitable and rice farms using ecological methods in the philippines in the largest set of adjacent farms totalling 1000 hectares using the biodynamic farming method, there was a yield increase of 50 to 100% and increase in the net income of farmers by 200 to 270% compared to the green revolution methods. Niconar Perlas a Filliphins agricultural scientist said while presenting these case studies that rapid transaction from chemical farming to sustainable agriculture was possible if correct technical principles are followed.

Cuba provides an outstanding example at a national level that has recently succeeded in moving from highly chemical intensive agriculture to ecology friendly approach.

Peter Rosset writes that in many cases peasent farmers had remembered old methods and re-applied them. In almost every case Rosset states they said that they had done two things remembered the old techniques like inter-cropping and manuring that their parents and grand parents had used before the advent of modern chemicals. Simultaneously incorporating bio-pesticides and bio-fertilizers into their production practices.

Jules Pretty had analyzed 45 non-chemical agricultural initiates spread across 17 African countries from these some 7,30,000 farming house holds have substantially improved food production and their food security. In 95% of the projects where yield increase was the aim, cereal yields have improved by – 50 to 100%. Total production has increased over all. An increasing number of people are seeking a way out of the problems created by chemical intensive agriculture by following the agro-chemical approach.

Miguel Altieri an agro-ecology expert at the university of calfornia recently estimated that there are already about 5 million hecters of farms being recuperted through ecological methods by two and half million families around the world.

Suggestions

1. Need for Govt. intervention in Motivating people to follow the organic farming.
2. Encouraging and facilitating farmers for exchanging their knowledge and experiences about the organic farming.
3. Educating the people about the importance of maintening ecological balance and bio-diversity etc.

Reference

Sustainable Agricultural Development in India book edited by R.S. Ramesh, K.S. Venkateshappa & S. Suryaprakash Southern Economics publication.

Article from Seminar monthly March 2009.

21

Watershed Strategy for Sustainable Agricultural Development

T. Gurubasappa, M.S. Kallur

During late sixties and early seventies, there was a major break through in Indian Agriculture with the green revolution and white revolution which placed the country in respectable position of self sufficiency in food grains and milk production. However, the tempo could not be maintained during the eighties. The uneven development due to green revolution has created socio-economic imbalance between the small farmers and large farmers and also regional imbalances in the country. The green revolution is largely confined to assured irrigated areas.

Further, Ecological imbalance is also caused by several factors both at macro level and regional level. No doubt, battle against hunger was won by the strategy of irrigated agriculture; it has not been achieved without huge environmental costs. Firstly, agricultural resources in several areas have been severally degraded. Signs of Agro-Eco System stress, and even of its break down, are abundant. Long ago Kanwar and Bali (i.e., 1977) estimated that 90 million hectares suffer from water erosion, 50 million hectares from wind erosion, 7 million hectares are effected by salinity and alkanity, and 20 million hectares from loading. In recent years, all these have been on the tremendous rise.

Secondly, sustainability of production in irrigated areas is not possible as there has been a substantial reduction in the storage capacity of big water reservoirs. On account of accelerating erosion, which has caused siltation of storage structures at much faster rate than what was assumed in their initial designs. We cannot, now, rely on irrigated agriculture for on the one hand, we have to increase our, food grains productions and on the other, per hectare yield of irrigated areas has already started declining. Hence, the realization of this food granis target is possible only if our rain fed crop lands of 148 m. ha. Area developed to their full potential.

Besides, the bloom of the green revolution, the rate of growth in agriculture has generally lagged behind overall growth of the national economy. The stagnation that set in during the decade and a half period beginning 1980, when the growth rate hovered around 3%, dropped to a mere 2% for the period 1995-2005; later this further nosedived to 1.5%. Yet the 11th five year plan envisages a 4% rate of growth in agriculture—a difficult mission by all counts.

The growth in productivity of most of the important food crops like rice, wheat, coarse, cereals and Oil seeds was negative for the period 1996-2004. When the food demand is projected to touch 240 million tonnes in 2010, production stood at less than 220 tonnes during the year 2007-08; India indeed is at the brink of loosing the status of being self-sufficient in food grain production.

Even as the contribution of agriculture to GDP has fallen from 37% in 1983 to around 19% currently, the share of workforce engaged in agriculture has hardly comedown.

Erratic and inadequate rain fall adversely affects the agricultural production in dry lands. Soil erosion is the serious problem on the productivity front. About 175m.ha. of the total geographical area suffers from various types of land degradation. Dry lands are becoming poorer day by day and this poses serious limitation on the moisture storage and sustainability of the crop productivity.

Realising the vital role and importance of rain fed farming in Indian economy and restoration of ecological and socio-economic balance, the Government of India has resolved that, "Ending neglect of vast rain fed and dry land areas" is a major policy issue and has adopted the watershed approach for integrated and comprehensive development of rain fed areas. The approach aims at scientific land-use through development of integrated rain fed farming systems on the principles of "WATERSHED" management in each development block where more than 70 per cent of the arable area is under rain fed condition.

Watershed management aims at minimizing risks associated with rain fed farming by following steps:

1. Conserving soil and water resources through mechanical and or cultural methods.
2. Draining out excess water at a safe velocity and directing it for safe storage for its utilization in dry season.

3. Preventing gully formation through mechanical and vegetative means and storage of water for recharging ground water.
4. Utilizing land according to its capability and putting marginal land unsuitable for arable crop production to alternate land use.
5. Developing a sustainable eco-system in harmony with the man-land-water-plant animal complex of the watershed.
6. Optimizing agricultural productivity per unit area, time and available water and
7. Improving the quality of life of the watershed inhabitants through infrastructure development.

The soil and moisture conservation programmes were initiated both at the State and Central levels in the country right from the first five year plan The emphasis was on small farm development and income distribution at the cost of Scientific conservation and management of resources. The integrated and sectoral approach to the problem in the early 1970's as drought relief measures reported to have contributed to accelerated soil erosion as a result of improper design, layout and maintenance of contour bunds and associated structures.

In drawbacks of such earlier approaches led to the emergence of watershed concept as one of the most scientific tools for resource conservation and development during 1980's. In fact, watershed intervention concept in the country was started about half a century ago (i.e. during 1940's). But the emphasis then was on protecting the down stream water resource infrastructure from the effects of upstream over exploitation of land. However, the focus has generally shifted towards in situ conservation and Management of the upstream to optimize and stabilize productivity of these areas on a sustainable basis.

A watershed is a geographical, ecological and socio-economic unit. Watershed development is a holistic approach, which includes the development of both agricultural as well as non-agricultural lands with in the watershed. In addition, it also takes care of effective utilization of rainfall, conservation of soil, moisture and biodiversity. Proper management of non-arable lands falling in the category of common property resources by producing fodder, fuel, fiber, timber and other needs of the masses within it, has the greatest advantage of meeting the various equity obligation on the one hand and restoring the ecological balance on the other. Watershed development is also a co-operative and institutional building approach as people have to 'act together' with respect of cropping pattern, management of non-arable land, soil and water conservation, etc. This may solve the various problems of 'externalities' and basic needs production within the watershed.

Karnataka is the pioneer state in the large scale implementation of watershed development programme, which started with launching of the prestigious World Bank funded Pilot project at Kabbalanala Watershed in 1984. At the same time, four Dryland Development Boards (DLDB) and 19 District Watershed Projects were started in the same year. A total of Rs.5225.12 million has been invested on watershed development in the state till march 1997 of which Rs.4674.22 million were invested after 1984. Available evidences suggest that these projects are successful in growing additional crops, enhancing stable yields and net returns of crops and creating more employment (Krishnappa 1994).

Government of India accepted watershed as a unit of area development on the recommendation of the Task Force on integrated Agricultural Development in Drought Prone Areas, which submitted its report in 1973. It was from the fifth five year plan that the Government of India strongly recommended the planning and implementation of the DPAP through, watershed approach instead of taking the whole district as a unit.

Integrated Watershed Programme as a Strategy for Sustainable Agricultural Development

In dry lands, limited availability of water is the most severe constraint on crop production. Conservation of as much rain water in situ as possible and harvesting and storage of the excess run off are essential for alleviating this constraint. This is best done when the watershed approach is adopted. After having duly recognized the need for promoting dry-land agriculture in the country, at last, Government of India, assigned a high priority to the watershed approach for development of dry lands in its Eighth Five Year Plan and since then the same policy is being continued. Government of Karnataka has taken the Watershed development programme very seriously from 1983. As a result, it has initiated watershed projects in all its rain-fed districts. All the 5 districts of Hyderabad Karnataka Region has witnessed the implementation of watershed projects resulting in the proper use of dry land and the problem of ecological imbalance has been tackled effectively on thousands of hectares of cultivable dry land. Althrough watershed projects can be developed by threee approaches namely (i) Bureaucratic approach, (ii) NGO approach, (iii) Participatory approach. The best among them is the third one and two watershed projects of HKR (a) One in Gulbarga district and another from (b) Bidar district in which participatory approach is implemented has here been taken up for an indepth analysis.

1. *Katangidda Nala Watershed Under Integrated Wasteland Development Project, (MYRADA) Gulbarga District:*

It is located 20 km North-East of Chincholi taluka and comprises of 10 villages and 12 Tandas. The village Shadipur is 20 Km. away from Chincholi beneficiaries of the study area are totally 180 and out of them 60 have been interviewed with a questionnaire. Water harvesting structures have been constructed with people's participation and there has been a step up in the number of wells constructed in the study area and the water level in them has increased considerably. People have accepted a better cropping pattern and per hectare yield of major crops has increased considerably (see Table 1 & 2). Cropping intensity has increased due to availability of assured water supply and farmers are raising two crops in a year rather than one (see Table 3). In the past, agricultural activities were limited and farmers used to migrate to other places in search of employment. As a result of implementation of the project which has led to double cropping and undertaking of more remunerative activities like horticulture, sericulture, etc., the migration has decreased considerably. (see Table 4).

Table 1

Changes in mixed cropping practices

Sl. No.	Before	After
1.	Sesamum	Sesamum + Pigeon pea
2.	Sorghum + Redgram	Hybrid Jawar + Pigeon pea
3.	Groundnut + Pigeon pea(local)	Groundnut + Pigeon pea (HYV)
4.	Greengram + Horsegram	Hybrid Bajra + Pigeon pea
5.	Blackgram + Groundnut	Blancgram + Pigeon pea

Source : Primary Data.

Table 2

Per Hectare yield of major crops

Sl. No.	Name of Crops	Per Hectare Yield		Additional	Percentage change
		Before	After		
1.	Redgram	4.00	7.00	3.00	75%
2.	Greengram	2.50	4.50	2.00	80%
3.	Blackgram	2.50	5.00	2.50	100%
4.	Bengalgram	4.00	6.50	2.50	62.50%
5.	Hy.Jawar	3.50	6.00	2.50	71.40%
6.	Sunflower	3.50	7.50	4.00	114.28%
7.	Sugarcane	3.50	6.00	2.50	71.40%

Source : Primary Data.

Table 3
Cropping Intensity of the study area

Sl. No.	Size of Land Holding (in Acres)	Total Area (in Acres)	Kharif Cropping Intensity	Rabi Cropping Intensity	Total Cropping Intensity
1.	0 to 2.5	26	100	38	138
2.	2.6 to 5.00	35	100	42	142
3.	5.01 to 10.00	58	100	47	147
4.	10.01 & above	90	100	51	151
	Total	**209**	**100**	**51**	**151**

Source : Primary Data.

Table 4
Migration Trend

Sl. No.	Size of Farmers	Before	After
1.	Marginal (20)	20	04
2.	Small (15)	15	12
3.	Medium (15)	15	-
4.	Large (10)	-	-
	Total	**50 (83.3%)**	**16 (26.6%)**

Source : Primary Data.

2. Participatory Watershed Planning of (PRAWARDA) Bidar District

Participatory watershed planning was implemented on cultivable area of 1880 hectrare belonging to 1133 families of seven villages. As in the case of Shadiput project, we selected purposively 60 beneficiaries of the project for eliciting detailed information. People dependent directly and indirectly on watershed resources were persuaded to organize into Self Help Groups (SHGs) and User Groups (UGs). In the study area one UG was formed for every 90 acres of land area. To carry out the Watershed planning and implementation at village level, specific institution called "Watershed Management Committees" (WMCs) were formed and their members were trained in participatory watershed planning, implementation and book-keeping (see Table 5). The chairman and secretary of

Table 5
Details of Watershed Development Project (WDP) of PRAWARDA:

Sl. No.	Name of the Village	Area Treated Under WDP	Name of WMC	No. of Members of WMC
1.	Sastapur	125	Kaveri Jalanayan Samiti	13
2.	Mirjapur	138	Ganga Jalanayan Samiti	15
3.	B.N.Wadi	120	Jalanayan Samiti	15
4.	H.Wadi	230	Jalanayan Samiti	14
5.	Koudiyal	116	Ganga Jalanayan Samiti	15
6.	Nirgudi	600	Jalanayan Samiti	15
7.	Illhal	557	Bramalya Jalanayan Samiti	16
	Total	**1880**		**103**

Source: Progress Report of Prawarda October 2003, March 2004, Sastapur, Basavakalyan, Bidar District.

the WMCs did play a key role in seeing that constructions like farm bonding, waste weirs, boulder checks, diversion channel, form ponds and agro horticulture are completed on time. We have come to know from them that they personally took measurements of structures, maintained records of payments and indulged in 'curing' activity.

Water harvesting structures completed on degraded upper reaches have raised the water level in open wells in the lower reaches. Open grazing has been controlled in all the seven villages. Agriculture related interventions in the form of introduction and use of improved varieties of green gram, pigeon pea, chick pea resulted in wholesome developments like reducing the crop period avoiding the problem of germination of seed while in pods and per acre yield improvements were from 26% to 100%. Respondents said that their lands were kept follow in the past and watershed structures have provided the required moisture and as such, they have started cultivating them. Improvement in the farming system has taken place and respondents have become familiar with what is called 'Integrated Farming System' where in different crops like Ginger, Chilly and

Tomato have increased their returns. As cropping intensity has increased, the migration has come down substantially.

Conclusion

Agriculture is and will remain the backbone of our rural economy. It will continue to employ the bulk of our population, as it should. All that is required in promotion of sustainable agriculture that is in harmony with nature, and desisting from policy prescriptions that depress farmer's incomes and force them out of farming.

Further, the experience of the region points out the reality that watershed development projects are the proper and feasible remedial measures for the ecologically degraded dry land areas. This is true of other regions of the country. Therefore, efforts are being continued at both micro and macro level to implement more and more watershed projects with participation of the people as a strategy for sustainable agriculture development.

References

Biradar A.A. (1991), Techno-economic issues of watershed development approach – A case study of district watershed, Gulbarga (Karnataka), *Indian Journal of Agriculture Economics*, 46 (3) 313.

Bhaskar Goswami, 'The Problem' Seminar Agrarian Transitions (595) a Symposium on the Growing Distress in Agriculture. March, 2009.

Chaurasia S.P. and Singh L.R. (1991), Optimal cropping pattern for minimizing soil loss in Naurar Watershed of Almora District (Uttar Pradesh). *Indian Journal of Agriculture Economics*. 46 (3) 320.

Ghosh M.G. (1991). An evaluation of the National Watershed Development programme in West Bengal, *Indian Journal of Agriculture Economics*, 46 (3) 299.

Hedge, B.R. (1989), Crop Land Development in Govt. of Karnataka, Watershed Development Programmes for Rainfed Agricultural, State Watershed Development Cell, Bangalore.

M.S. Kallur; Dry Farming, Prasaranga, Gulbarga University, Gulbarga 1998, p. 16.

Micheal A.M., "Stress on Water Management" *Survey of Indian Agriculture*, 1994, The Hindu, pp. 4-8.

R.G. Desai, Watershed Management – A Study Development of Dryland Agriculture Problems and Prospects (2004) book edited by: Dr. Rasure Khemaling.

Profile in soil conservation, Dry Farming Research Station, Sholapur, Golden Jubilee Year, 1983, p. 38.

22

Deforestation and Its Impact

S. Nalinakshi

Deforestation

Natural Resources are the resources available in nature, which are useful to man, and it can be transformed into useful product.

Forest resources are the renewable resources on this Earth:

1. It provides a variety of highly valuable Ecological Economic and Social Services.
2. Forest conserves biological diversity, soil and water.
3. Forest enhances agricultural production improves urban living.
4. Forest provides a wide range of wood and non-wood products.
5. Forest performs some ecological services like
   (a) Production of oxygen.
   (b) Reduction of global warming,
   (c) Wildlife habitat.
6. Forest regulates hydrological cycle.
7. Forest prevents soil erosion and moderates pollution.
8. Forest provides variety of fruits leaves seeds roots which be used for food and medicines.
9. Human beings in the development are responsible for deforestation.

Deforestation means removal of vegetation or damage of vegetation in the forest. It is permanent destruction of indigenous forest and woodlands. Indigenous forest now covers 21% of the Earths land surface.

Causes for Deforestation

Increasing of Human Population

Human population is mainly responsible for conversion of forest and woodlands to agricultural land to feed growing number of people. To compound this problem the poor soils in the humid tropics do not support agriculture for long. Thus people are forced to move on and clear more forest in order to maintain production.

Shifting Cultivation

In attempt to settle farmland the poor became shifted cultivators and resort to using slash and Shifting Cultivation - burn method of tree removal. Slashing and burning involves cutting of trees and remains are burned. The ash is used as fertilizer and the land is used for farming and cattle grazing.

Fuel Requirement

Increasing demands for fuel wood by the growing population in India alone has shooed up in recent years thereby increasing the pressure on forests. Fuel wood collection can contribute total depletion of tree cover particularly in rural and hill areas.

Raw Materials for Industrial Use

Wood is important raw material for many purposes. Ex For making boxes, furniture, railway sleepers, plywood, matchbox, and pulp for paper industry have exerted tremendous pressure on forest. Development projects like construction of Dams, hydroelectric projects; road construction, etc., cause deforestation. Forest fires also causes lot damage to the forest.

Mining

Mineral deposits are found in the forest area and any operation of mining will naturally affect the forest. Deforestation in Missouri and Deharadun due to mining over a length of 50,000 hectares of forestland. Mining of radioactive minerals in Kerala, Tamil Nadu, and Karnataka are posing threat of deforestation.

Effects of Deforestation

Global Warming

The most serious consequences is that of climate change due to loss of forest Earth has an atmosphere which contains variety of gases, one of is carbon dioxide, a gas which helps moderate heat loss to outer space. Insulating gases such as Co_2 are called Green houses gases because their function is like that of the glass in the greenhouse; they allow solar heat into the system but discourages its escape. Other green house gases include methane, chlorofluorocarbons, nitrous oxide and ozone. These gases increase the temperature on the earth surface. This leads to changes in weather patterns, sea level and other cycles in nature directly affects the life on Earth.

Destruction of Forest Resources, Wildlife Species and their Natural Habitat

Deforestation is causing a loss of biological diversity on an unprecedented scale. Although tropical forest covers only six percent of Earths land surface, they happen to contain 70% &90% of all of the world's species. As a result of deforestation, we are losing between 50&100 animal and plant species each day. Inevitably, the loss of species entails a loss of genetic resources.

Many of these species now facing the possibility of extinction is of enormous potential to humans in many areas; especially medicine. As of 1991, over 25% of the worlds pharmaceutical products were from tropical plants. By contributing to the extinction of multiple species of plants & animals, we might be destroying the cures of many of diseases that diseases that plague the human race today. The world's forests, particularly tropical rainforests, are home to over 10million members of the "last surviving intimately resources-based cultures". Given the importance of forest products to daily lives of forest people's destruction of forest entails the destruction of tribal population as a whole. Aboriginal people worldwide have had their land literally stolen by government and industries whose intent is to turn natural capital into hard currency. As the Global Futures Foundation states, there has been more extinction of tribal peoples in this century. Even in the rare cases when forest dwellers are compensated for this loss the changes visited upon their culture by the inexorable of industrial culture are devasting. Without a doubt, deforestation hashed propounds effect on cultural diversity throughout the forest region and ultimately, the world.

Hydrological Cycle and Rainfall

Flooding is quite serious consequence of deforestation. Clearing of forest dramatically increases the surface run-off from rainfall, mainly because a greater proportion of the rain reaches the ground due to lack of vegetation, which would suck up the excess rainfall. Tropical forests can receive as much rain in an hour as London would expect in a wet month, and a single storm has been measured as removing 185 tonnes of topsoil per hectare. In tropical regions where the forests are dense, flooding is not as serious a problem because there vegetation to absorbs the rainfall. It is areas where there is little vegetation that there is a problem. Hence to avoid the disastrous effects of flooding need to remain dense & lush.

Soil Erosion, Loss of Fertility and Landslides

The lushness of world's tropical forest is somewhat deceptive. Although these forest assume to be lush and full, the under lying soils are very poor, almost all the nutrients being bound up in the vegetation. The problem is that once the forest is cut down, essential nutrients are washed out of soil all together. This leads to soil erosion. As of now about 80%soil in the humid tropics are acidic and infertile. When there are no trees to keep the soil becomes ripe for erosion.

Impact of Deforestation

Desertification in Hilly Regions of the Himalaya

Deforestation in Himalayas, involving clearance fo natural forests and plantation of monocultures like *Pinus roxburghi, Eucalyptus camadulensis* etc. have upset the ecosystem by changing various soil (edaphic) and biological properties. Nutrient cycling has become poor, original rich germplasm is lost and the area is invaded by exotic weeds. These areas are not able to recover and losing their fertility. The entire west Khasi hill district of Meghalaya in North-east Himalayas, Ladakh and parts of Kumaon and Garhwal are now facing the serious problem of desertification.

Desertification of Khasi Hill District of Meghalaya due to Deforestation

One of the world's wettest areas is suffering from a shortage of water. The Khasi hills, in a remote part of north–east India, usually experience torrential rains. The area once recorded more than 1000 inches of rain in just year – a world record. But villagers in the region, which is named after rain filled clouds that supplied the waterfalls and streams, now have to bring water from other areas. Increases in pollution and deforestation have blamed for environmental changes.

Disappearing Tea Gardens in Chhota Nagpur

This hilly region used to be a good frosted area towards the turn of the century

Khasi Hills - Villagers has to carry water from the plains

In the monsoon, Cherrapunjee's waterfalls become raging torrents

A COAL MINE close to the town. The slopes of the Khasi hills have been stripped bare by logging and mining.

and used to receive fairly frequent afternoon showers favoring tea plantations. Following the destruction of forests, rainfall declined in Chhota Nagpur to such an extent that tea gardens also disappeared from the region.

Warning Rainfall in Udhagamandalam (Ooty)

The sub normal rainfall during 1965-84 at Ooty in Nilgiri Mountains has been found to be closely associated with declining forest cover in this region in the past 25 years. The rainfall pattern was found to fluctuate with wooded land area in the hills. When the Nilgiri Mountains had luxuriant forest cover annual rainfall used to be much higher.

Forest Cover is Decreasing in Andaman and Nicobar

Deforestation and legal rights to forest resources on the island of Little Andaman, in the Andaman and Nicobar Island group, an Indian territory located in the Bay of Bengal. The island is the home of the Onge—a threatened indigenous community of Negrito origin. The island was clothed in tropical evergreen forests and inhabited only by the Onge through the middle of the 20th Century. In the 1960s, the Government of India introduced major development and colonization programs for the island. These programs have destroyed vast areas of forest and have severely affected the Onge. Over the last 55 years, roughly 40% of the island has been taken over by outsiders for settlements, agriculture, timber extraction, and plantations. These developments have adversely affected the island's flora and fauna as well as the Onge themselves. Logging operators and government agencies have systematically violated both the laws and the resources of Little Andaman. A coalition of Indian social and environmental non-governmental organizations (NGOs) initiated legal action to stop the deforestation in 1999. These efforts led to landmark injunctions on all timber felling operations in the islands by the Supreme Court of India in October 2001 and May 2002.

Measures of Deforestation

Tree Plantation

There is no cure for deforestation. But many people talk of reforestation, however that is not a true solution. Although replanting the forest that has been destroyed seems like a good idea, it actually does no good. Often times the new trees are not the same species as the originals. Also by time the trees grow and mature, the soil has already lost much of the nutrients it once had. Old forest and new forest are not the same and it is the old forest that needs to be protected.

Strict Implementation of Law of Forest Conservation

1. Lack of vision, in foreseeing environmental problems, not evolving appropriate policies, plans and programmes, besides non-dynamic, reactive (rather than being, pro-active), legislative laws, in tackling the complex and ever challenging environmental issues and problems appear to be at the root of the activist stance of the courts of law.
2. The entire process of evolving policy and law, in the country is a flawed one. It is both an illogical and absurd process. We have policies without laws, laws without policies and policies following legislative efforts!
   (a) The National Agricultural Policy and those of the States, is a classic example of a policy without law. No single legislation encapsulates the contents and concerns of the policy document. This has to be gleaned from a variety of bits and pieces of legislations. Agricultural processes, production, marketing, support price, etc; have not attracted the attention of the lawmaker as yet.
   (b) Quite a few environmental legislations do not having the backing of a policy document. The Wildlife (Protection) Act, 1972, The Forest (Conservation) Act, 1980; Water (Prevention and Control of Pollution) Act, 1974; The Water (Cess) Act, 1977 and Air (Prevention and Control of Pollution) Act, 1981, are only a few examples of such "stand alone" documents.
   (c) We have the classic case of putting the cart before the horse, with the framework Environment (Protection) Act, 1986, preceding the National Policy and Strategy for Environmental Protection and Sustainable Development. The latter following the former by a gap of 6 years!

Education and Awareness Program

We the human beings may not understand severity of the possible consequences that deforestation poses. Since deforestation has had no severe effect on us yet, we ignore the problem. Everywhere you go, you see pieces of paper on the ground, people using multiple tissues to wipe their noses, and countless people pulling excess amounts of brown paper out of the paper towel dispensers in lavatories. These are just few of the sources of paper that we use each day, without any thought whatsoever.

What we must realize is that the paper products we use daily is a part of a forest which functioned to enrich and hold soil, absorb CO_2, collect and recycle water, release oxygen and regulate climate. Some companies do plant trees to produce the kinds of products needed by the industries to spare the older forest but many do not. By wasting paper, we are wasting the forests. The simple fact is that the more paper we use more forest need to be cut down to serve our paper needs.

Many people might not consider the possible consequences of deforestation serious. They might say: What if a few people lose their home. We experience little flooding here and there; the temperature rises bit; we may miss out few medicines; we kill off a few species; the soil may loses its nutrients. Ignorant people like these do not realize the severity cf these consequences.

The only way to ensure that we will not encounter any of these consequences of deforestation is to stop destroying the forests all together. We would have to stop cutting down all trees, no matter what our needs were. Since this is totally impractical as of now, the only thing we can do is a use forest product in moderation. However this idea of moderation needs to be put into practice immediately. We as people should take care of this problem before it gets to the point where we can no longer fix it. Instead of putting the problem on the back burner until we can no longer ignore it, how about we remedy it now so it never threatens us to begin with. The sad fact is that once the forest is gone, we wont be able to fix the damage, which we caused.

Conclusion

There is no one easy solution as deforestation is caused by many things. One option is decreasing the need for the amount of products that are harvested from the rainforests. If all countries, especially developed ones, enforced programs that used recycling, the need for disposable products would be diminished and the loggers would not have a business. If the demand is cut off, there is no need for the supply.

The immediate effects of deforestation may not yet be felt, but if this generation doesn't feel it the next generation and their children will be the ones to suffer. It is the actions of the human race that can make or break the future of the planet. We the human beings should adopt the concept of ahimsa ensure the protection and conservation of all forms of life there by keeping the ecological balance of earth intact.

Some important Earth ethics are:

1. Love and honour the Earth that provides us life.
2. Keep each day sacred to Earth.
3. Do not waste the resources.
4. Avoid polluting the environment.
5. Do not hold yourself above the other living things and have no rights to drive them to extinction.

References

Perspectives in Environmental Studies, by Anubha Kaushik & C.P. Kaushik, New Age International Publishers.

Tropical Deforestation: *The Human Dimension,* by L.E. Sponsel, T.N. Headland, R.C. Bailey.

Forests, by Renu Khator, Publisher: South Asia Books.

Himalayan Perceptions: Environmental change and the well-being of mountain peoples, by Jack D. Ives.

Applied Geography: *Principles & Practice,* by Michael Pacione.

The Earth as Transformed by Human Action, by Billie Lee Turner.

Deforesting the Earth: From Prehistory to Global Crisis—An Abridgment, by Michael Williams.

Tropical Forest Ecology: The Basis for Conservation and Management, by Florencia Montagnini & Carl F. Jordan.

23

Use of Biogas in Imphal District

Mamatha, B. Jennifer Serto and
K. Banadarangaiah

Introduction

Energy is a prime mover of economic growth. But, due to increase in energy prices and fast depletion of fossil fuels coupled with environment pollution, deforestation, soil degradation, global warming and so on made us to think of alternative energy technologies like solar, wind, bioenergy that are available abundantly and free from environmental implications.

India with its huge bovine population has an abundant supply of animal dung. Its value as manure and as a source of fuel has been known to the farmers for centuries. However, a given quantity of dung could be put only to one use that is as manure in agricultural fields or as cooking fuel in the form of dung cakes. It was only from the last sixty years or so that the biogas technology has enabled a dual use of this resource, as a source of fuel and as a rich organic manure and in turn helps in the sustained growth of economy and contributes for the all round development of our country. Hence this study was undertaken with a purpose of eliciting information on the use of biogas in Imphal a district of Manipur state. The objectives of the study are

1. To find out the usage of biogas and raw material used for biogas plant.
2. To ascertain the benefits of biogas.

Methodology

To collect information on the use of biogas, survey method and questionnaire was the tool used. A sample of sixty household was selected using purposive random sampling technique. A pilot study was conducted to test the feasibility of the questionnaire. A modified questionnaire was used for the main study. The questionnaire consisted of questions related to background information of the respondents, use of biogas and slurry, source of raw material, maintenance, financial assistance, problems and benefits on the use of biogas. The data collected were compiled, analyzed statistically, tabulated and interpreted.

Results and Discussion

The background information of the respondents reveals that most of them were in the age group from 31 to 40 years and the income ranged from Rs. 5000 to Rs. 10,000 per month. Higher percentage of the respondents belonged to nuclear family. The educational qualification varied from illiterate to graduate and post graduate level.

Majority of the respondents used biogas as a major source of fuel for cooking and heating water and a few for lighting. Firewood was used as a secondary fuel for cooking, charcoal for heating water and kerosene for lighting.

All the respondents possessed Deenbandhu fixed type biogas plant. The respondents opined that the gas produced by biogas plant was sufficient for their family. They received information from relatives and friends, nodal agencies and only a few percentage through newspapers and magazins. The cost of biogas plant ranged from Rs. 10,000 toRs. 15,000.

Table 1 reveals that 93 per cent of the respondents used biogas for cooking, 77 per cent for heating water and a few of them used for lighting purpose. Further 42 per cent of the respondents used slurry as an organic manure for gardening and 31 per cent for agriculture and 27 per cent used slurry for both gardening and agriculture. Therefore this table shows that biogas is not only used for cooking and heating water but also for lighting and residue as rich organic manure.

Table 1

Source of raw material used for biogas plant:

Source of raw material	Percentage
Animal dung	100
Animal urine	67
Kitchen waste	05
Human waste	50
Crop residue	04

Table 2 depicts that all the respondents used animal dung followed by animal urine and human waste as raw material for their biogas plant. A few of them used kitchen waste and crop residue.

Table 2

Use of biogas and Slurry

Use of biogas	Percentage
Cooking	93
Heating water	77
Lighting	28
Slurry	
Manure for agriculture	31
Gardening	42
Both	27

Maintenance of Biogas Plant

Higher percentage (77%) of the respondents did not hire any labour and the cost was nil for maintaining their biogas plant. But 23 per cent of them hired labour and the labour charges ranged between Rs.1000 and Rs. 2000 per year.

Financial Assistance

Majority (82%) of the respondents availed financial assistance from government and a few of them spent their own money to install biogas plant. The respondents who availed subsidy was in the range of Rs. 6000 to Rs.10,000.

Fig. 1 shows that most of the respondents opined that they did not face any problem in using biogas plant but 32 per cent opined that there was crack in the construction and 18 per cent had the difficulty in repairing biogas plant. A small percentage of the respondents said that there was a lack of facilities for the repair of biogas plant.

Table 3 reveals that higher percentage of the respondents opined that the use of biogas was economical followed by environment friendly, easy to maintain. A small percentage of them opined that it helps to bring down fuel crisis and also it is hygienic.

Conclusion

Biogas is a high energy fuel that can be used as a substitute for fossil fuel and a rich source of organic fertilizer. As a wide range of organic substance from agriculture, kitchen waste, animal and human waste can be used to produce gas

Table 3

Opinion on the benefits of the use of biogas

Benefits	Percentage
Economical	62
Easy to maintain	45
Helps to bring down fuel crisis	26
Environment friendly	51
Hygienic	20

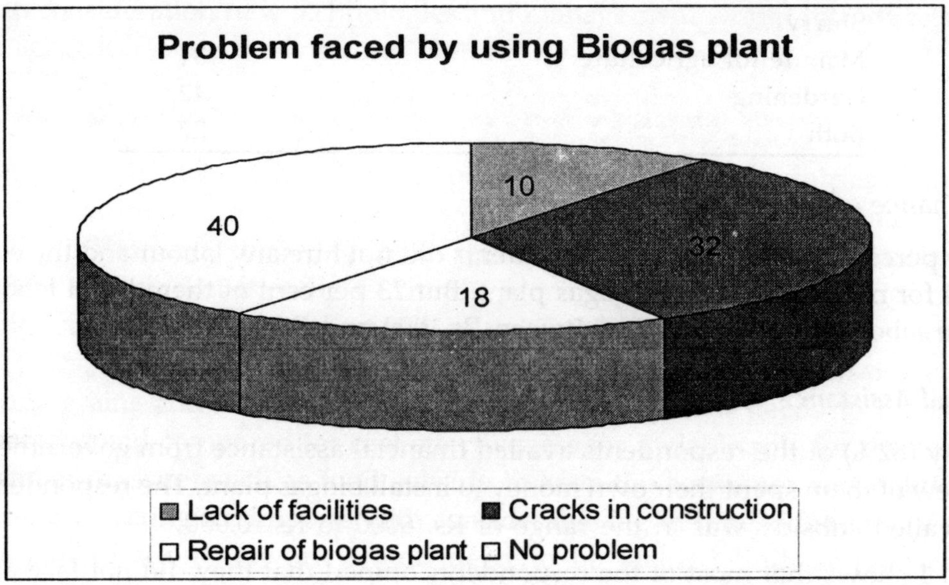

Fig. 1

and it is one of the promising technologies for waste management. It also makes a significant contribution in preventing drudgery of women, deforestation, soil degradation, global warming and an indispensable element together with wind, water and solar power of today's renewable energy mix.

References

G.D. Rai (1996) " Non – Conventional energy " 4th edition , Khanna publishers, Nai Sarak Delhi.

www.google.com

24

Intensive Irrigation Induced Poor Land Productivity in Upper Krishna Canal Commands—An Analysis

V.B. Kuligod, G.N. Kulkarni and S.T. Hundekar

Indian burgeoning population demands more, nutritious and balanced food. It has been estimated that in India 211 million people go to bed hungry each day. India would be food deficient to the extent of 50 mt by the year 2030 AD. This is not an unachievable task in near future because the country possesses advanced crop production technology and capable of adopting appropriate planning at national and regional levels. Creation of irrigation facility, timely and sufficient availability of inputs, development of high yielding hybrids and verities, mechanization and efficient pest and disease management are few of the interventions done at various levels to meet the food requirement during green revolution era. However, these modern production technologies with improper adoption have lead to stagnation or decreased food productivity in the recent fast especially in the irrigation commands. Soil salinity, alkalinity, waterlogging, deteriorated soil health, incidence of pest and diseases, violation of recommended cropping pattern, imbalanced fertilizer application and many others are the causes that led to lower productivity compared to targets. According to one estimate, an area of about 1.41 M ha (about 10.5 per cent of the existing command area) has been suffering from varying degrees of waterlogging and salinity.

Keeping this in view, the paper discusses sustainable land, crop and water management in UKP canal commands with scientific information and ground realities.

Upper Krishna Project (UKP) Command Area

The UKP is the prestigious and an on-going major irrigation project of Karnataka in the drought prone Gulbarga, Raichur, Bagalkot and Bijapur districts. On its full development, it would command an area of 1 M ha. The project area lies between $75^{0}15'$ and $77^{0}30'$ E longitudes and $15^{0}45'$ and $17^{0}30'$ N latitudes. The overall efficiency of the irrigation system has been about 50 per cent. The climate is semi-arid tropical characterized by short monsoon, mild winter and hot summer. The weighted average annual rainfall is 701 mm, while the average annual evaporation is 2,493 mm (based as the long-term weighted average of Gulbarga, Bijapur and Raichur data). The monsoon usually sets in early June and lasts till the end of September. The dry season starts from mid-February and continues through the middle of June. April-May constitute the hottest period. Winter season is from November to mid-February. The temperature regime in the command area is hypothermic.

The command area has highly diversified geological formations comprising of peninsular complex - granite, granite–gneiss, limestone, shale, sandstone and deccan trap (basalt) rocks. There are three distinct landscapes-granite and deccan trap landscapes consist of undulating uplands to nearly level plain while the limestone landscape mainly consists of gently sloping midlands. The area has a good network of streams and the drainage pattern is dendritic. The rivers Krishna and tributaries Bheema and Don constitute chiefly the drainage system of the command area. The soils are typical of semi-arid areas of Central India and broadly consist of two types viz., black soils constitut ing 86 per cent (deep soils-57 per cent; shallow to medium deep soils-29 per cent) and the red soils cover the remaining 14 per cent.

Studies on the analysis of Krishna water indicate that its quality is good for irrigation and is classified as C_1S_1. According to Groundwater Survey Investigation Wing, UKP Command Area Development Authority, in Stage-I Phase-I area, the area underlain by brackish water is 26,703 ha. The investigations on water quality further showed that the EC was within 2 dS/m in 63,534 ha and in the range of 2-2.5 dS/m in 1,08,450 ha. Generally, in red soils the ground water is fit for irrigation, while in black soil areas it is of inferior quality due to high SAR and RSC values (Subramany, 1989). The recent studies indicate that the quality of ground water in Stage-I Command area is good and can either be used directly or mixed with the canal water. Only in an area of 6,014 ha in Narayanapur Left Branch Canal (NLBC) and Shahapur Branch Canal (SBC), the groundwater quality is unsuitable for irrigation even by mixing with water (Anonymous, 1997).

Why Land Productivity and Sustenance Jeopardized in UKP

Cropping Pattern and Yields

Sustainability and higher productivity are jeopardized in the UKP due to inappropriate cropping pattern, imbalanced and unscientific commercial fertilizer use in addition to poor irrigation water management. Cropping pattern before and few years after irrigation introduction was dominated by the sorghum, *bajra,* red gram, green gram, groundnut, chick pea and sunflower. These crops require less irrigation water and, they were sustainable and profitable for long years of traditional irrigated and dry land agriculture. However, the cropping pattern got changed due to ample and unregulated availability of water since 1995. The high nutrient and water requiring commercial and food crops like hybrid cotton, chillies, sugarcane and paddy dominated cropping pattern in the irrigated situation of the UKP. This altered not only soil fertility status but even the agricultural ecosystem and made agriculture unsustainable. Semi-arid climate, deep black soils with poor internal permeability, undulating topography and improper land development for irrigation doesn't make suitable conditions for cultivation of crops like paddy. The dominance of cropping pattern by paddy, chillies and cotton has lead to problems like:

1. Higher incidence of pests and diseases to unmanageable levels.
2. Deteriorating soil fertility, productivity and general soil health.
3. Extensive soil salinity, alkalinity and waterlogging.
4. Luxurious growth of exotic and native weeds in the lands and natural drains.
5. Seepage and surface run-off resulted in low water use efficiency

Balakrishnan *et al.* (1999) have estimated that the farmers are realizing 60, 52, 46, 41 and 38 per cent lower yields compared to potential yields in crops like cotton, sorghum, sunflower, maize and pearl millet, respectively, in Upper Krishna Project (UKP) command area. Deteriorating soil physical and chemical conditions, depleting plant nutrients and wide spread occurrence of waterlogged saline lands are the main causes for lower yields in addition to poor pest management. Therefore, there is an urgent need for crop, land use and irrigation water planning to make irrigated agriculture sustainable in the UKP command area.

Problems of Waterlogging and Salinity

After the advent of irrigation, the water table in the Stage-I Phase-I command area is being monitored by the Ground Water Survey and Investigation Wing of

UKP CADA (Command Area Development Authority) using observation wells on 5 × 5 km grid basis since 1984 for NLBC command and since 1987 for SBC command. The Department of Mines and Geology, Government of Karnataka also has been monitoring the same since 1981 (using 17 wells). The monitoring revealed that the water table is increasing at alarming rate in the UKP command area. The water table observation data recorded by the Mining and Geology Department for Stage-I Phase-I area indicate that during 1984 – 1995, the average annual rise in water table was 6 cm. According to Ground Water Survey and Investigation Wing of UKP CADA during 1984 – 1990 in Stage-I Phase-I area, the water table rose @ 13 cm per year. The waterlogged area has gone up by nearly three times from 3,200 ha (<1 m bgl (below ground level)) in 1984 to 9,400 ha during 1995. The areas likely to be waterlogged with water table at 1-2 m bgl were estimated at 34,150 ha. The number of villages affected by waterlogging (with water table <2 m bgl) has been increasing over the years; it has from 25 villages in 1988, increased to 38 in 1981, to 56 in 1994 and to 80 in 1995 as per the monitoring surveys. During November 1995, 23 villages had water table <1 m and 57 villages had water table in 1-2 m range.

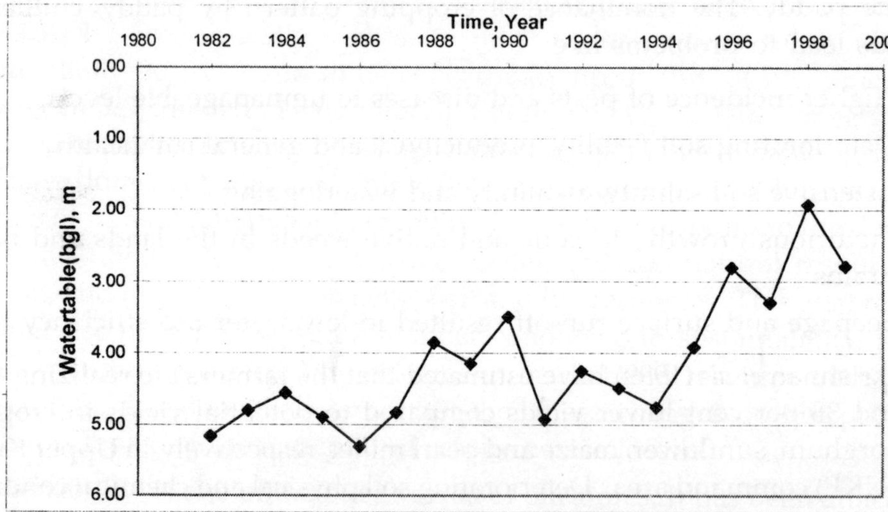

Fig. 1. Average rise of watertable after introduction of irrigation in UKP.

The salts derived from parent rock start accumulating in the soil profiles of the command area due to inefficient leaching on account of scarcity of rainfall. Of the 30 soil series observed in the command area, 18 series are found to have alkaline tendency. It has been found that 60 per cent of the soils in the canal commands of the Stage-I have salinity concentration below 3,000 me, 30-35 per cent area has 3,000–6,000 me and the remaining 5-10 per cent area has

6,000-10,000 me. It is possible that with the advent of irrigation, the areas having salinity concentration upto 6,000 me[2] would get leached in course of time that may result in salinity concentration favourable to crop production. However, the same introduction of irrigation has rendered considerable area as waterlogged and salt-affected in NLBC and SBC commands over the year due to water table rise. These areas need to be reclaimed through appropriate drainage measures and amendments if any depending on the nature of the problems.

The problem of salinity/sodicity already existed in the Stage-I area prior to irrigation due to aridity of the climate. The pre-irrigation surveys indicate canal command wise, that an area of 26,226 ha was salt-affected in Gulbarga and Bijapur districts. The statistics show that 4.28 per cent of the total area surveyed was affected by salinity/sodicity when no canal water was introduced. The salinization in the irrigated command was mainly due to secondary salinization induced by canal irrigation. Due to advent of irrigation in NLBC and SBC commands during 1983–'84 and 1986-87 respectively, an area of 45,000 ha out of 46,792 ha in NLBC command and 75,000 ha out of 1,22,437 ha in SBC command were brought under irrigated agriculture by 1994–95. During this period the problems of waterlogging and salinity manifested in about 947 ha in NLBC and 1,271 ha in SBC commands. Thus, during the period 1983–84 to 1994–95, based on the water table monitoring as discussed elsewhere, an area of 11,618 ha (about 2 per cent of the irrigated lands) was rendered waterlogged

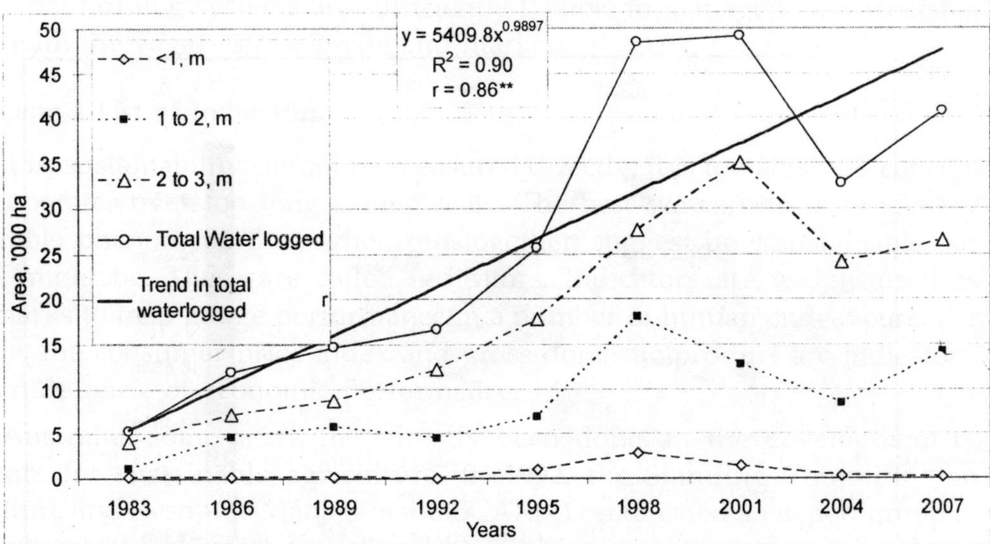

Fig. 2. Area under different categories of the waterlogging during last 2.5 decades in the NLBC and SBC canal commands of UKP during *rabi*.

and salt-affected. However, since so systematic surveys have been carried out, and considering the nature of the command and the existing irrigation practices, the actual figures might be still higher.

The recent investigations by Kulkarni et al., (2008) have revealed that area under waterlogging (with water table within 3 m) substantially increased to 26,969 ha in 2006-07 from a meager 1,046 ha of before irrigation introduction during *kharif*. This phenomenal increase was within a period of 2.5 decades of canal irrigation in the NLBC and SBC commands. Corresponding increase was from 650 ha to 44, 612 ha during *rabi* season. Compound growth rate in total water logged area during *kharif* and *rabi* seasons were 6 and 10 per cent, respectively.

Another recent survey carried out in shorapur *taluka* by Kuligod (2003) using inductive electromagnetic meter indicated that surface soil salinity (0-15 cm) was as high as 28 dS m^{-1}. High salinity were observed all along the NLBC canal running from Golbal village to Kembhavi (> 12 dS m^{-1}). Similar magnitudes of salinity were observed in and around Kakkera and Peth Ammapur villages and along the road connecting Kembhavi and Shorapur. However, non-saline patches were also recorded near Rajankollur and Narayanapur villages as the soils are of red type characterized with good drainage. Saline patches were predominant in black soils regions (Kakkara, Hunasagi, Kembhavi and Peth Ammapur villages) especially in the areas where farmers practice paddy cultivation. The problem of salinity was prevalent in about 0.40 lakh ha. The salt

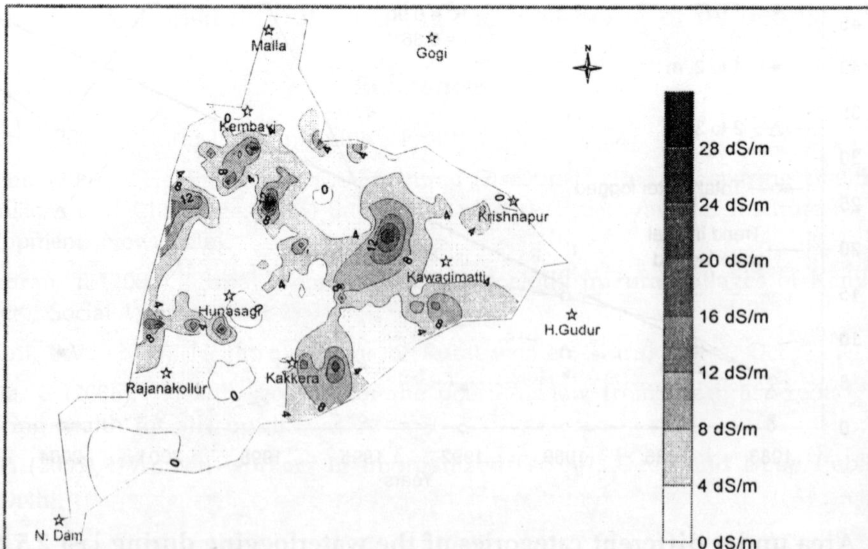

Fig. Soil salinity in shorapur taluk during summer 2002 (0-15 cm depth)

patches are extensively seen in low-lying areas due to undulating topography. The soil salinity, though not serious, was observed even in uplands (elevated fields).

The various causes of waterlogging and salinity in the command area are poorly drainable black soils, seepage from canal network, lack of land development, unscientific irrigation practices and inadequate drainage. The quality of Krishna water (0.6 dS/m) and that of the groundwater available in 80 per cent of the UKP Stage-I command area (<2 dS/m) were found good for irrigation. The estimates of salts imported by irrigation of canal and ground waters to various crops varied from 0.7 to 25.3 t/ha/season (Table 1). The groundwater available for exploitation annually has been assessed at 377 MCM (million cubic metre).

Though the waterlogging and salinity problems pose serious threat to agricultural lands and the environment, the drainage measures have not got the required attention. Only surface field drains of 1-1.5 cusec (0.057-0.085 cumec) capacity were excavated in an area of 25,120 ha till July 1998, while the attempts on subsurface drains (SSDs) were sporadic. The Master Drainage Plan for UKP Stage-I identified 84,215 ha for SSDs. The land development works were carried out in about 1/4th of the notified irrigation development area only.

Looking to the enormity and seriousness of the problems of waterlogging and salinity, the present plans seem to be grossly inadequate. Therefore, mitigative measures are urgently required to halt the increasing trend and reclaim the

Table 1

Salts added to soil through irrigation per season

Sl.No.		Crop m/season)	Irrigation (t/ha)	Yield	Salts imported (t/ha)
				Canal Water (EC=0.6 dS/m)	Ground Water (EC=2.5 dS/m)
1.	*Cotton*	700	2.5	2.2	10.9
2.	*K. paddy*	1225	5.4	3.8	19.0
3.	*R/S. paddy*	1625	6.1	5.0	25.3
4.	*R. sorghum*	450	2.8	1.4	7.1
5.	*Wheat*	400	2.2	1.2	6.2
6.	*K. groundnut*	325	1.1	1.0	5.0
7.	*R. groundnut*	650	2.0	2.0	10.2
8.	*Bajra*	225	2.2	0.7	3.5
9.	*Pigeonpea*	600	1.2	1.9	9.3
10.	*K. sunflower*	325	1.4	1.0	5.0
11.	*R. sunflower*	475	1.3	1.5	7.4

K - Kharif., R/S- Rabi/Summer.

affected areas. Under the circumstances, the results of Indo-Dutch Network Project would go a long way in serving the UKP command area in prevention/ reclamation of waterlogged saline areas. This would help in preservation, restoration and sustainability of the agricultural land production capabilities.

Deteriorating Soil Fertility

Growing of high yielding verities and hybrids and increased cropping intensity has led to depletion of the soil fertility status. Further the farmers tend to apply more of nitrogenous fertilizers. This has caused imbalance in the nutrient content of the soils. A survey has indicated that only 2 per cent of the land is sufficient in available nitrogen. Phosphorus is most deficient in the command. About 96 per cent of the area is deficient in available P. Vijayasekhar et. al. have reported that black soils of UKP are not only deficient in major nutrients but even in the important micronutrients like zinc and iron. Zinc was deficient in 91-100 per cent of the samples that had been analyzed. Another micronutrient, iron was deficient in 25-53 per cent of the samples. Therefore, there is an urgent need for educating the farmers on judicious and balanced use of the chemical fertilizers and integrated plant nutrient management to make soil more productive and sustain the productivity.

Farmers of the command are ignorant on balanced and judicious fertilizers use. In comparison to other commands, average consumption remain one of the lowest and it is skewed towards more application of nitrogenous fertilizers with gross neglect of micro and secondary nutrients.

Poor Water Use Efficiency

On-farm development in respect of land leveling, bunding and maintenance of irrigation channels is either partial or non existence. Large undulations in the topography were found to occur within and between the fields. This coupled with large size of the fields and steep slope in the upper reaches hampered irrigation water management. A preliminary studies of irrigation distribution network under use revealed that the FICs were in bad shape without proper sections and/or without slabs in many fields. This is attributed to poor quality construction, weed growth and poor maintenance. The discharge in the FICs varied widely and the average seepage losses were found to be 1.93 and 1.71 cumec/M m^2 in cases of FICs in banking and cutting respectively for an average discharge of 18 lps; the seepage losses ranged from 3.1 to 3.51 cumec/M m^2 in case of damaged FICs.

Farmers practiced appropriate irrigation methods for various crops. Common irrigation method in cropped lands remained wild flooding till today.

The stream size used in irrigation water application varied from 1.2 to 3.0 lps in furrow irrigation. Generally, the land holding size ranged from 0.5 to 14.0 ha. The average longitudinal slope of the fields was in the range of 1.23 to 1.50 per cent except for paddy lands and the cross (transverse) slope varied from 0.3 to 0.5 per cent. The inventory clearly revealed that lack of land development and unscientific irrigation practices added to the low water use efficiency and higher seepage and surface losses. The field irrigation application efficiency is about 80 per cent while the overall efficiency of the irrigation system has been less than 50 per cent.

The land, water and crop productivity being low in UKP command, need of the hour is to enhance the productivity by adoption of the appropriate strategy and plans. The road map can have following components.

1. Massive programmes on land shaping.
2. Short and along-term land use panning.
3. Adoption of appropriate extension strategy.
4. Educating and training of farmers on efficient water use.
5. Participatory water management by active participation of farmer's societies.
6. Large scale land drainage programmes.
7. Extensive programmes on Sustainable soil fertility management.

References

Balakrishanan, P.; Kulkarni, G.N. and Kuligod, V.B. (1999). A status report of UKP in Karnataka. *Tech. Report*, IDNP, ARS, B'gudi.

Kulkarni, G. N., Kuligod, Vittal and Bharamagoudar, T. D. (2008). Intensive irrigation induced waterlogging in the Upper Krishna Canal command. The paper presented in the International symposia on Natural Resource Management in Agriculture held at Agricultural Research Station, RAU, Durgapura, Jaipur between 19-20, December, 2008.

Vijayasekhar, K., Kuligod, V.B. Basavaraj, P. K. Dasog, G.S. and Salimath, S.B. (2000). Studies on the micronutrient status in the important black soil series of UKP command, Karnataka. *Andhra Agril. Journal*, 47: 141-143

Subramany, M. (1989). Preliminary appraisal of the ground water quality and scope for conjunctive use in the Stage-I and Phase-I command area, Gulbarga District, Karnataka, *Ground Water Studies*, No.1, UKP-CADA, Bheemarayanagudi Karnataka.

25

Sustainable Agriculture through Self-help Group Approach

Muniswamy, S. Raju. G. Teggelli, Yamanura, G. Girish, S.M. Matapati

Introduction

Sustainable Agriculture through group approach found more effective (Nagaraj,N.2003). A self help group (SHG) has been defined by RBI as a small, economically homogenous and cohesive group of rural poor voluntarily forming a group to save small amount regularly, agree to contribute to a common fund, meet their emergency needs on mutual help basis, make collective decisions, solve conflicts through collective leadership and provide collateral free loans in terms of conditions defined by groups. Thus self help groups can ensure empowerment of rural people through their initiative being facilitated by a specified official support. The SHGs have now become synonym for participatory rural development. Sustainable income generation is given more importance than supply of consumptive loans and members are made aware about utility of savings. Farmers are group of rural people who are mostly exploited in various levels of farming operation. Participatory technology development measures are undertaken and supported by trainings on topics selected by group members regarding field problems that were unsolved before. This further results in profitable farming. With this background a study was conducted to know comparative performance of individual farmers and self help group farmer members.

Methodology

Two varietal demonstrations one each on Sorghum and Bengalgram was given totally to 40 farmers. Out of which 20 farmers belongs to 2 SHG's (Siddeshwara sanga and Arunodaya sanga) in village Kodli, Tq. Chincholi, Gulbarga (Dist.) and remaining 20 individual farmers belongs to 2 villages (Srichand and ladmugali, Tq.Aland). Preseason training was given to all members and critical inputs were supplied to them. One off campus training was conducted in each village during crop growth period and farmers are trained on crop management aspects. Field visits were made during critical periods of crop growth and recorded observations. Finaly field day was conducted and collected feed back from farmers. Based on feed back information came to know about difference in individual demonstrations and SHG involved demonstration. It was clear from the feed back that there was difference in the two demonstrations conducted in individual farmers fields and SHGs members fields.

Findings and Discussion

The findings of adoption of recommended technology results and feed back information are presented under the following major headings, viz., results of demonstration and timely field operations carried out.

From the above results it was found that though percent increase in Bengalgram FLD is same in both individual and SHG involved FLD, average yields were higher in SHG involved demo. Because of constructive and positive interaction of group. Similarly with respect to Sorghum FLD per cent increase in SHG demonstration is less (7.3 %) compared to individual demonstration. But average yields are higher in SHG demonstrations.

Average yields of Bengalgram and sorghum was higher in SHG demonstrations mainly because members of SHG share their knowledge and experiences during monthly meetings. They dicuss problems and arrive at conclusions. Hence they could achieve more yield than individual demonstrations.

Table 1
Results of demonstration

Sl. No.	Crop	Individual Demo Yield (q/ha)				SHG Demo Yield (q/ha)			
		High	Low	Average	Check	High	Low	Average	Check
1	Bengalgram	16.5	10.2	14.5	12.6	18.5	12.0	15.6	14.0
2	sorghum	17.8	11.0	15.2	13	19.0	12.5	16.1	15.0
	% increase over check (Bengalgram)-11.5%					% increase over check (Bengalgram)-11.5 %			
	% increase over check (Sorghum)- 11.6%					% increase over check (Sorghum)- 7.3%			

From the above Table-2 we can conclude that timely field operations carried out by SHG members with regard to sowing, seed treatment , IPM and harvest

Table 2
Timely field operations carried out

Sl. No.	Crop	Bengalgram- FLD		Sorghum- FLD	
		Individual	SHG	Induvidual	SHG
1.	Right time of sowing & synchrony	85% (17)	90% (18)	85% (17)	90% (18)
2.	Seed treatment	90% (18)	100%(20)	95% (18)	100%(20)
3.	Adoption of IPM practice	85% (17)	100%(20)	80% (16)	95% (19)
4.	Right time of harvest & synchrony	90% (18)	95% (19)	90% (18)	95% (19)
	MEAN %	87.5 %	96.25 %	87.5 %	95 %

Values in the parenthesis indicate number of farmers

was more synchronized and at the appropriate time. The mean per cent of timely operations was 87.5 % both in Bengalgram and sorghum individual FLD. As compared to SHG demonstration it was 96.25 and 95 per cent respectively in Bengalgram and sorghum FLD. SHG members gave feed back on fodder quality of soghum, varietal response to lodging, pest& disease reaction, grain yield, roti making quality of sorghum and drought tolerance capacity where as individual farmers gave feed back only on fodder yield and grain yield.

Regarding Bengalgram FLD, SHG members gave feed back on grain yield, response to nutrients(biofertilizers, FYM, & chemical fertilizers) pest and disease reaction, where as individual FLD farmers gave feed back only on grain yield. Feed back given by the SHG was more detailed and serve as source for further research.

Conclusions

The findings of the study indicate that effective control of pest and disease and right decision regarding agronomic practices, viz., sowing time, inter cultivation, spray schedule, harvesting etc., was observed in SHG. Since members of SHG meet often, they discuss field problems and agricultural activities to be carried out. The feed back given by the group was more valid and serve as better ingredient for further research. Thus group approach like self help group or farmer field school are more effective in achieving goals of sustainable Agriculture.

References

Ann R. Braun, Graham Thiele and María Fernández., (2000). Farmer Field Schools and Local Agricultural Research Committees: Complementary Platforms for Integrated Decision-making in Sustainable Agriculture.

Nagaraj, N. (2003). Farmer Field School in Karnataka Community Based Tank Management Project Implementation Guideline, Community Based Tank Management Project Consultancy Series.

26

Response of Bidi Tobacco to Zinc Sulphate

P.S. Matiwade, S.T. Kajjidoni and
A.R. Hundekar

Introduction

Bidi tobacco is an important commercial crop of Northern Karnataka. The bidi tobacco growing soils are showing the deficiency symptoms of zinc but they are hidden. This is due to continues use of straight fertilizers and no organic matter. There is need to study the role of zinc in the production of quality tobacco. Hence, the investigation was conducted.

Materials and Methods

The experiment was conducted in a randomized block design with four replications (plot size 9m × 6m) for three years from 2002-03 to 2004-05 at Agricultural Research Station, Nipani, Karnataka. The experiment was having eight treatments consisting of application of zinc sulphate alone and in combination with FYM and RDF for bidi tobacco. Different delivery methods were also followed for zinc sources such as basal application and spray at 0.2 and 0.4%. The data on growth parameters and yield were recorded for individual years and data was analysed statistically for individual years and for pooled data as well.

Results and Discussion

The pooled results of three years (2002-03 to 2004-05) and economics are presented in Table-1, 2 and 3.

Table 1

Plant height, No.of leaves, Leaf length and leaf width of bidi tobacco as influenced by zinc sulphate application (Pooled of three years - 2002-03 to 2004-05)

Tr. No.	Treatment Details	Plant Height (cm)	No. of Leaves/ Plant	Leaf Length (cm)	Leaf Width (cm)
1	RDF without FYM and no zinc	80.2	12.0	50.0	24.5
2	RDF with FYM 10 t/ha and no zinc	86.3	11.8	52.0	24.7
3	RDF + FYM 10 t/ha + 10 kg/ha $ZnSO_4$ once at planting	88.3	12.8	53.1	26.7
4	RDF + FYM 10 t/ha + 20 kg/ha $ZnSO_4$ once at planting	87.2	12.8	54.8	28.4
5	RDF + FYM 10 t/ha + 2 kg/ha $ZnSO_4$ for nursery	88.1	12.2	52.8	26.4
6	RDF + FYM 10 t/ha + 4 kg/ha $ZnSO_4$ for nursery	87.2	11.9	52.5	25.9
7	RDF + FYM 10 t/ha + 0.2% $ZnSO_4$ spray at 35 DAP	86.7	11.7	54.0	27.1
8	RDF + FYM 10 t/ha + 0.4% $ZnSO_4$ spray at 45 DAP	87.2	11.8	53.2	25.9
	S.Em ±	1.4	0.2	0.7	0.5
	C.D. 5%	4.0	0.6	2.0	1.5

Table-2

Bidi tobacco yield as influenced by zinc sulphate application

Tr. No.	Treatment Details	Yield (kg/ha)			
		2002-03	2003-04	2004-05	Pooled
1	RDF without FYM and no zinc	1267	712	1225	1068
2	RDF with FYM 10 t/ha and no zinc	1292	754	1217	1087
3	RDF + FYM 10 t/ha + 10 kg/ha $ZnSO_4$ once at planting	1517	787	1408	1237
4	RDF + FYM 10 t/ha + 20 kg/ha $ZnSO_4$ once at planting	1583	879	1442	1301
5	RDF + FYM 10 t/ha + 2 kg/ha $ZnSO_4$ for nursery	1371	825	1350	1182
6	RDF + FYM 10 t/ha + 4 kg/ha $ZnSO_4$ for nursery	1437	683	1346	1156
7	RDF + FYM 10 t/ha + 0.2% $ZnSO_4$ spray at 35 DAP	1325	795	1333	1151
8	RDF + FYM 10 t/ha + 0.4% $ZnSO_4$ spray at 45 DAP	1354	745	1296	1132
	S.Em ±	59.5	35.94	72	29.6
	C.D. 5%	174.9	105.73	NS	87.1

Table-3
Economics of bidi tobacco as influenced by zinc sulphate application

Tr. No.	Treatment Details	Gross Income (Rs/ha)	Cost of Cultivation (Rs/ha)	Net Returns (Rs/ha)	B : C Ratio
1	RDF without FYM and no zinc	29670	17250	12420	1.72
2	RDF with FYM 10 t/ha and no zinc	30690	18750	11940	1.64
3	RDF + FYM 10 t/ha + 10 kg/ha $ZnSO_4$ once at planting	34560	19060	15500	1.81
4	RDF + FYM 10 t/ha + 20 kg/ha $ZnSO_4$ once at planting	36930	19370	17560	1.91
5	RDF + FYM 10 t/ha + 2 kg/ha $ZnSO_4$ for nursery	32940	18812	14128	1.75
6	RDF + FYM 10 t/ha + 4 kg/ha $ZnSO_4$ for nursery	31800	18874	12926	1.68
7	RDF + FYM 10 t/ha + 0.2% $ZnSO_4$ spray at 35 DAP	31800	19500	12300	1.63
8	RDF + FYM 10 t/ha + 0.4% $ZnSO_4$ spray at 45 DAP	31470	20100	11370	1.57

D:\My Documents\Agronomy\PSM article.doc

Price of Tobacco Rs. 30/kg
Price of $ZnSO_4$ Rs. 31/kg

Yield of bidi tobacco was significantly influenced by various treatments of $ZnSO_4$ during 2002-03, 2003-04 and in pooled analysis. In pooled analysis, the application of 20 kg $ZnSO_4$ once at planting along with RDF and FYM recorded cured leaf yield of 1301 kg/ha and was significantly higher than that obtained from the other treatment combinations except the treatment wherein 10kg $ZnSO_4$/ha once applied at planting along with RDF and FYM (1237 kg/ha). During 2002-03, application of zinc sulphate @ 20 kg/ha along with RDF and FYM 10 t/ha recorded maximum cured leaf yield of 1583 kg/ha followed by zinc sulphate 10 kg/ha (1517 kg/ha). However, spraying with zinc sulphate @ 0.2% and 0.4% did not record significant yield improvement. The untreated check (RDF without FYM + no zinc) recorded minimum yield of 1267 kg/ha. Similar trend was noticed during 2003-04 and 2004-05. Application of $ZnSO_4$ as INM component also supported the growth parameters such as plant height, number of leaves, leaf length and leaf width. The zinc sulphate application @ 20 kg/ha supported growth parameters and thus influenced in recording the maximum yield, Sarma et al. (2005), reported 17.6% higher cured leaf yield due to application of 20 kg $ZnSo_4$/ha. Similar results were reported by Patel et al. (2005), Vageesh et al. (2005), and Shetty et al. (2005).

Conclusion

It can be concluded that the application of zinc sulphate @ 20 kg/ha has significantly influenced the productivity of bidi tobacco and be recommended as useful component in INM of bidi tobacco in Karnataka.

References

Patel, J.R. Ghelani, L.M. Hirpara, B.V. Patel, S.B. and Shah, S.N., 2005, Integrated nutrient management in bidi tobacco. Paper presented in the *National Conference on Tobacco: Scientific Strategies for Sustainable growth of farm economy and export,* Hyderabad, 3-7th October, 2005.

Sarma, Y.N., Bisoi, B., Mahapatra, H.K. and Prasad Rao, J.A.V., 2005, Effect of zinc sulphate on growth yield and quality of pikka tobacco. *Paper presented in the National conference on Tobacco: Scientific Strategies for Sustainable growth of farm economy and export,* Hyderabad, 3-7th October, 2005.

Shetty, Y.V. Rudramurthy, H.V., Vasuki, N., Vagesh, T.S. and Veerabhadrappa, B.H., 2005, Effect of zinc and magnesium on yield and quality of FCV tobacco. *Paper presented in the National conference on Tobacco: Scientific Strategies for Sustainable growth of farm economy and export,* Hyderabad, 3-7th October, 2005.

Bageesh, T.S., Nataraju, S.P., Rudramurthy, H.V. and Vasuki, N., 2005, Influence of zinc on leaf yield and quality of FCV tobacco. *Paper presented in the National conference on Tobacco: Scientific Strategies for Sustainable growth of farm economy and export,* Hyderabad, 3-7th October, 2005.

27

Changing Cropping Pattern in India

Gachinamani Monappa

Introduction

Desirable changes in cropping pattern of a country in the farm diversification of crops would cater to the needs of industry in more than one way. Firstly, it would provide the necessary wage goods for the secondary sector. Secondly, it would encourage agro-based industries like sugar, cotton and jute and there by help industrialization of the economy. It is particularly true of a predominantly agricultural economy like India. "India is the world's largest producer of Sugarcane, sugarcane processing including the production of mile, sugar, khandsari and gur, is a major industry".

A cropping pattern which give scope for growing of several crops the products of which are in great demand in foreign countries also serves the export sector of its economy. For instance, "many of the commercial crops loom large in India's foreign trade. India is a traditional exporter of oilseeds cakes and meals; particularly deoiled groundnut and cotton seed meals, of speciality oils such as caster and linseed; and of hand picked, selected groundnuts. Jute is one of the important agricultural exports and it was the largest exchange earner in the early seventies. Today, India is not only surplus in raw cotton but also produces one of the finest varieties like suvin. The country now has comparative advantage in the production of staple cotton and many Indian varieties are rated amongst to the finest in the international market".

The cropping pattern is so important for predominantly agricultural country, in particular, and to majority of the countries, in general depends on several factors. "Soil fertility, its texture, topography, rainfall, climate, water availability, consumers tastes and preferences, social structure, agro-industrial set up, infrastructure, agro-industrial set up, infrastructure and over all agrarian conditions are some of the determinants of cropping pattern. Moreover, economic forces such as expected prices, credit availability, marketability etc., also alter the cropping system of a region".

Objectives

1. To study the Changing Cropping Pattern in India
2. To make suitable suggestions.

What is a Crop?

According to Encyclopaedia Britanica, "Crops are plants and their products usually grown with special care or for a special purpose and in most instances harvested to yield a product for which there is a market demand. Emphasis is placed on the ability to produce a harvest".

Meaning of the Cropping Pattern

The cropping pattern is defined in the following paragraphs: 'cropping pattern means most efficient use of land and other resources over a time. It is, therefore, necessary that the suggested cropping pattern offer some alternative plans for the farmers to maximize production per unit area per unit time. In our concept, an efficient cropping pattern must ensure greatest efficiency inputs. Thus, the cropping pattern means both the time and space sequence of crops'.

"Commonly, crop pattern refers to the allocation of area to different crop activities. It is a process usually determined by the interaction of natural and man-made resources of any region in the long run. Soil fertility, its texture, topography, rainfall, climate, water availability, consumers tastes, agro-industrial set up, infrastructure and overall agrarian conditions are some of the determinants of cropping pattern".

"Crop pattern is mean the proportion of area under different crop at a point of time. A change in cropping pattern implies a change in the proportion of area under different crops".

"Under crop pattern, we study the area of land devoted to different crops, the changes in it overtime, and the factors that determine the division of land under different crops".

In the light of above definitions, we may define the cropping pattern as the proportion of area under different crops at a point of time and it is determined by factors like fertility of the soil, climate, topography, rainfall, water availability, consumers tastes, etc.

Factors Determining the Cropping Pattern

1. **Physical and Technical Factors:** Cropping pattern in any region depends upon physical characteristics as the soil, climatic weather, rainfall, etc. Apart from soil and climatic conditions, the cropping pattern of a region will depend upon the nature and availability of irrigation facilities. Wherever water is available, not only a different crop can be grown but also double or triple cropping will be possible. Physical compulsion also determines the cropping pattern of a given region. For instance, water-logging of an area would lead to an increase in the paddy area because rice can stand the extra water better than other crops. In newly reclaimed lands, millets are to be grown for the first three to five years before the soil is ready to shifted to other superior crops like rice. It is possible that because of lack of capital, agricultural perquisites, better implements, improved seeds and finance for getting fertilizers, it might not have been the right crop that was being grown, but given these facilities, the cropping pattern may change.
2. **Historical Factors**: At any time, crop pattern of any country is decided by history. For instance, 'the most ancient system of cultivation seems to have been what is called shifting cultivation which survives in many parts of the world, including India, among primitive people'. There was nothing like systematic cropping patter. Now, the rotation of crops on scientific basis is experienced by both developed and developing countries. This has been made possible by the application of science and technology to the primary sector in these countries.

   The way in which tenurial systems have been developed over a period of time is also an important factor in shaping the crop pattern. A small farmer naturally gives importance for foodgrains production for his consumption purpose. Landlords always prefer non-food crops.
3. **Economic Factors**: Economic motivations are the most important factors in determining the cropping pattern of a country. However, they are explained below:

   (a) ***Price and Income Maximisation***: Farmers are very much influenced in the choice of their crops by considerations such as (i)Price parities between different commodities, (ii) Maximisation of income, (iii) relative profitability per acre. The prices influence the acreage under the crops in two ways. One is that the variations in the inter-crop price parities lead to shifts in acreage as between the crops. Another is that the maintenance of a stable level of prices for a crop which provides a better incentive to the producers to increase the output than what a very high level of price does, if there is no uncertainty of this level being maintained over a number of years. The 'income maximization pull' has greater influence in changing the crop pattern of a farmer. In other words, the farmer would choose that combination of crops which would give him maximum income.

(b) ***Farm size***: There is a definite relationship between farm size and cropping pattern. The small farmers are first interested in producing foodgrains for their requirements. In other words, usually the need for subsistence dominates the cropping of the small farmer. They would go in for cash crops only after they have met their requirements of foodgrains. However, now-a-days, the traditional relationship between farm size and cropping pattern has changed substantially.

(c) ***Insurance against risk***: The need to minimize the risk of crop failure not only explains diversification but also some specific features of crop pattern. For instance, the persistence of millets in many regions which puzzles many economists can be understood mainly as insurance against bad seasons in dry areas.

(d) ***Availability of inputs***: As already indicated, crop pattern is also dependant upon the availability of such inputs as seeds, fertilizers, water, storage and marketing, transport etc. Further, among all inputs, the most rewarding is 'irrigation' and it is now-a-days more important than others. Further, given normal weather conditions, transport and market factors also decide the income from crop production. So, there is an urgent need for the provision of efficient transport and well developed market centers for profitable and balanced cropping pattern. Finally, availability of labour also determines the cropping pattern of a region. For example, labour surplus induces farmers to in for short duration crops. On the contrary, labour shortage creates a preference for long duration crops.

4. **Government Action and Crop Pattern**: The Government can influence the crop pattern through legislative and administrative measures. Steps may be taken by the Government to ease or subsidise the surplus of the farm inputs and knowledge. The provision of irrigation facilities or the supply of seeds and fertilizers, etc., may be related to the adoption of a given crop pattern by the farmers. Government has to give the greatest

attention to provide adequate financial and other resources to the farmers, as factors like recurrent drought or pest-infestation may be preventing them from opting for a more remunerative set of crops. In such situations if more of irrigation, fertilizers, pesticides etc., are made available, then, it would be possible for them to change their crop structure and so earn larger revenue from their land.

5. **Social Factors**: Social factors which include density of population, customs, traditions, attitudes toward material things also influence the type of crop grown and the area devoted to different crops.

Crop Pattern in India

In olden days shifting cultivation was widely practices in our country as in other parts of the world. Gradually the shifting cultivation system gave way to the fallow system of cultivation. Later, with the development of use of animal dung as manure and the different ways of storing water for irrigation, the various field systems of cultivation appeared. As years rolled by, domestication of cattle and other farm animals progressed meanwhile and in course of time a pattern of mixed farming got developed through the ages which is practiced by the Indian peasantry, in general. Now, the average farmer has a piece of land to raise crops and a few heads of livestock and sometimes a few ducks or hens for meeting his various needs. Here an attempt is made to analyse the cropping pattern in India.

Crops in India

Fortunately, the country has a large variety of crops to its credit. The large extent of its area and a great variety of physical and natural features and climate combined with natural fertility of the soil, enable the country to produce almost every kind of vegetable life. Herein one finds not only the tropical products but also the products of the temperate zone as well. Our foodgrain crops include, rabi cereals like wheat, barley; kharif cereals like rice, maize, jawar, bajra, ragi, and other small millets, and various kinds of pulses like redgram, greengram, horsegram, blackgram etc.; commercial crops include oilseed crops, sugarcrops, tobacco, cotton, jute and other bast fibres. Horticulture crops grown in the county are mango, banana, potato, garlic etc. A large number of plantation crops are also grown in the country. The most important crops from the viewpoint of exports are tea, cashew-nut, coffee, pepper and coconut. Finally, we have fodder crops, particularly legumes which are important to feed the enormous livestock population of our country. Food grain crops provide the staple food to the country's population. Pulses are the main sources of protein to the vast vegetarian population of the country. Commercial crops provide the raw materials to the agro-based industries and they also earn a sizeable amount of foreign exchange. Horticulture crops are the important source of vitamins and minerals. These also supplement the carbo-hydrate and proteins to varying degrees depending upon their composition. Plantation crops, like commercial crops, are important one from the view point of export. Fodder crops are essential for the development of animal industry in India.

Main features of the Cropping Pattern

There are two important features namely: (1) a large variety of crops and (2) the preponderance of food crops over non-food crops.

1. A large variety of crops: In fact, the striking feature of the cropping pattern of India is its amazing variety. For instance, according to India 1981 the number of important crops was 23 and out of that 11 were grouped under foodgrain and the rest 12 were put under non-foodgrain crops. Further, tea, coffee, rubber, bananas etc., were neither included under non-food crops nor under foodgrain crops.
2. Preponderance of Food crops over non-food crops: The other peculiar feature of the cropping system in India is that the food crops dominate the scene than the non-food crops. For example, the three-fourths of the total cropped area is under food crops. The most important reason for this dominance is that the food is the basic need of life. Among food crops, the rice is the most important crop occupying roughly about 25 per cent of total cropped area. There are more than 7000 varieties used or cultivated, which occur in India. The next single crop in order of importance is wheat. Now-a-days, millets like, jawar, bajra, maize etc., are also becoming popular.

Trends in Cropping Pattern

At the beginning of the present century more than 83 per cent of land was under food crops and about 17 per cent under non-food crops. By 1944-45, area under food crops had come down to 80 per cent and area under non-food crops had increased to 20 per cent. Further, changes in the cropping pattern during the various years are mentioned in Table 1.

There has been decrease in the area of cereals (i.e. from 66.99 per cent to 59.52 per cent during the period from 1970-71 to 2000-01. As against foodcrops, area under commercials crops has increased from 18.18 per cent in 1970-71 to 28.49 per cent in 2000-01. Whereas, pulses show a decreasing trend during the same period.

However, Table 1 does not provide a detailed analysis of the trends of individual cereal crops, pulses and commercials crops. This detailed information is provided in Tables 2 and 3.

Trends in Individual Cereal Crops: The data shown in Table 2 provide information about individual cereal crops in India. The area under wheat crop has increased very impressively. For instance, the area under this crop has increased from 20.45 million hectares to 25.70 million hectares between 1970-71 to 2000-01. There is an increase of rice area from 37.47 million hectares in 1970-71 to 44.70 million hectares in 2000-01. The area under bajra has decreased over the years. For instance, bajra acreage has come down from 11.57 million hectares in 1970-71 to 9.80 million hectares in 2000-01.

Trends in Commercial Crops: As data presented in Table 3 among commercial crops, oilseeds have been growing faster than the other crops. Not only they are being raised, but also their growth is steady. It is also true of the sugarcane crop. Further, there is increase in area under sugarcane crops from 2.61 million hectares in 1970-71 to 4.30 million hectares in 2000-01. However, there is decrease in area under groundnut crop during the same period.

Suggestions for Improving the Cropping Pattern System in India

1. Providing Financial Assistance.
2. Ensure adequate and quality inputs like improved seeds (High yielding variety seeds HYV) Fertilizers, Pesticides, Tools, etc.,
3. Necessary transport, marketing & other infrastructure facilities should be provided to the formers.

Table 1

Area under different crops in India

(In million hectares)

Years	Cereals	Pulses	Commercial Crops	Total Cropped Area
1970-71	101.78 (66.99)	22.53 (14.83)	27.63 (18.18)	151.94 (100.00)
1980-81	104.21 (63.81)	22.46 (13.75)	36.65 (22.44)	163.31 (100.00)
1990-91	103.21 (58.10)	24.70 (13.91)	49.71 (27.99)	177.61 (100.00)
2000-01	100.71 (59.52)	20.30 (11.99)	48.20 (28.49)	169.21 (100.00)

Source: Economic Survey, Government of India, 2004-05.

Note: Figures in brackets denotes percentages.

Table 2

Area under Principal Cereals in India

(In million hectares)

Name of the Crops	1970-71	1980-81	1990-91	2000-01
Rice	37.47 (36.81)	40.15 (38.53)	42.70 (41.38)	44.70 (44.39)
Jawar	16.09 (15.81)	15.80 (15.16)	14.40 (13.95)	9.90 (9.83)
Bajra	11.57 (11.36)	11.65 (11.18)	10.50 (10.17)	9.80 (9.73)
Maize	6.03 (5.92)	6.00 (5.76)	5.9 (5.71)	6.6 (6.55)
Wheat	20.45 (20.09)	22.27 (21.38)	24.20 (23.45)	25.70 (25.52)
Other Cereals	10.19 (10.01)	8.33 (7.99)	5.50 (5.34)	4.00 (3.98)
Total Cereals	101.80 (100.00)	104.20 (100.00)	103.20 (100.00)	100.70 (100.00)

Source : Economic Survey, Government of India, 2004-05.

Note : Figures in brackets denotes percentages.

Table 3

Area Under Principal Commercial Corps in India

(In million hectares)

Name of the Crops	1970-71	1980-81	1990-91	2000-01
Groundnut	7.31 (18.89)	6.80 (16.71)	8.30 (16.70)	6.6 (13.69)
Oilseeds	16.64 (42.79)	17.60 (43.24)	24.10 (48.48)	22.80 (47.30)
Sugarcane	2.61 (6.71)	2.66 (6.53)	5.70 (11.47)	4.30 (8.92)
Cotton	7.60 (19.54)	7.82 (19.21)	7.40 (14.89)	8.50 (17.63)
Jute	0.70 (1.70)	0.94 (2.32)	1.00 (2.00)	1.00 (2.08)
Other Commercial Crops	4.03 (10.36)	4.88 (11.99)	3.21 (6.46)	5.00 (10.38)
Total Commercial crops	38.89 (100.00)	40.70 (100.00)	49.71 (100.00)	48.20 (100.00)

Source : Economic Survey, Government of India, 2004-05.

Note : Figures in brackets denotes percentages.

4. Farmers education programmer should be given top priority with regard to water resource management, developments in the field of technology market etc.
5. Farmers should be advised to grow suitable crops. Which will increase them productivity & income level during this reforms & WTO era.

Conclusion

The present cropping pattern tilted heavily in favour of food grains and it is not an conductive for a country. So there is a need for changing in existing cropping pattern so that the commercial crops can find their legitimate place in it.

Farmers must also change their outlook & their cultivation methods according to changing scenario of our economy.

References

Memoria C.B. Agricultural Problems of India, Kitab Mahal, Allahabad 1985.

Hussain Masjid "Crop Production in India" Concept Publishing Company, New Delhi 1982.

A Ramalingam.C. "Some Economic Aspects of Cropping Pattern"

Acharya S.S. "Crop Diversification in Indian Agriculture" Agriculture Situation in India August 2003.

Govt. of India, Ministry of Agriculture, Irrigation Commission-II, New Delhi 1972.

Economic Survey, Govt. of India, 2004-05.

28

Urbanization and Sustainable Agricultural Development

Revanasiddeshwara, M. Kuntimath
Jaganath, M. Lagashetty

Introduction

Most of the world population growth in the coming decades will be in the cities of developing countries. These cities will therefore account for most of the increased demand for food. Urban populations, however, are not only consumers but also producers of food, especially high-value perishable agricultural produce. The growing urban populations and their demand for food and other agricultural products will also have major impacts on land use in peri-urban areas and nearby rural areas. The present addendum examines the impacts of urbanization on sustainable agriculture, including the issues of urban agriculture, the impact of cities on pen urban areas and rural-urban linkages.

A majority of the population in North America and Europe have lived in cities since the early twentieth century, a majority in Latin America since the I960 and a majority of people in West Asia since 1980. People in Asia and Africa remain predominantly rural but that situation is changing rapidly. Asian cities are growing at an average rate of 3 per cent per year, compared to overall population growth of 1.4 per cent, while the population of African cities is growing at a rate of 4 per cent per year compared to overall growth of 2.6 per cent. In both regions the majority of the population will live in urban areas by 2020. By 2025, approximately 60 per cent of the world's population will live in urban areas.

The urban population living in poverty has rapidly increased during the past two decades, along with rapid urbanization. The United Nations Centre for Human

Settlements (Habitat) reported that in 1990, at least 600 million people in urban areas in Latin America, Asia and Africa were living in life—and health-threatening conditions. The additional 600 million people who would be added to urban populations in developing countries during the 1990 would increase the pressure on land, housing, water and sanitation, food supply and distribution system, and education and health services, and would add to the urban population living in poverty.

A Brief Review

Although most people in Western cities have lost their links with the land, this is not the case for the people who are now streaming into third world cities. While many of these rural refugees may not like the idea of continuing to work the land, the fact remains that they do have the survival skills to produce food on their own, given access to the necessary resources.

In two poor neighborhoods of Bombay, Pan Walker (1986) found that rural migrants who came to the city to find a job considered farming to be a demeaning activity that they had left behind. This attitudinal barrier is further complicated many African countries where the traditional role of women in subsistence farming reinforces the reluctance of men to engage in non-commercial food production.

Deelstra (1987) noted, however, that such survival skills could also be used to facilitate the transition to urban life. By producing some of their own food, immigrants can gain more confidence in themselves, leading to increased security and better integration. Urban agriculture also contributes to increasing one's social contacts, a prerequisite for success in the city.

This, in fact, appears to have been the case at least in Zaire, where Streiffeler (1987) observed that men adapted all too quickly to city ways. When it became evident that urban agriculture could be used for intensive, lucrative production of vegetables as cash crops, the attitudinal barriers vanished overnight and the men displaced the women. Although commercial ventures are traditionally a masculine role, these men generally had little experience in growing vegetables and thus required training courses. One may assume that the women had to grow their own crops on poorer or more distant plots.

Gutman (1987) concluded that 100 sq m of intensively cultivated land in Argentina could supply the vegetable needs of a family of five and that this would require about I to *1.5* working days per week. A successful garden in Buenos Aires could thus save between 10 to 30 per cent if not 40 per cent of the cost of an appropriate diet for a family, representing savings of *5* to 20 per cent or more of its total income.

In one of the few publications in this field, Browning (1985) made the point that home gardens produce not just vegetable often plants that provide starch, fruits, herbs, flowers, and medicines as well as fuel wood. In many cases, small livestock are also raised with their wastes helping to fertilize the garden whose wastes, in turn, help to feed the livestock. Young (1987) noted that although urban forestry for fruit production is rare in Asia, 25 per cent of the 'large number" of street trees grown in the Indian city of Bangalore bears fruit.

The economic value of home gardens has received some attention but most of it has dealt with the straight market value of the produce, which is difficult to assess in those areas where most food is self-produced or exchanged. While most people normally save money by growing their own food, this is particularly important in the third world where the food budget accounts for a much higher proportion of total family expenditures: often 50 per cent and reaching 70 per cent for urban families in India, compared with 25 to 30 per cent for the average American family. Another consideration is that with gardens ensuring part of a family's subsistence, there is a greater chance that people will risk investing in other income-generating activities.

Unlike capital-and energy-intensive agricultural development projects, urban agriculture does not depend on big budgets. It must be made quite clear, however, that urban self-reliance does not mean less work for the state or international agencies but a different kind of work. What is needed is the development of policies that support such initiatives rather than institutional barriers that block them. This is a crucial issue that must lead to new types of dialogue between communities and local authorities.

Sanyal (1986) noted that the evolution of government positions on urban agriculture—from outright rejection, through benign neglect, to lukewarm and occasionally enthusiastic encouragement—will face stiff resistance from the many bureaucrats who played major roles in shaping the industrialization strategies of their countries. These "modernization" experts and their supporters generally oppose any efforts perceived as threatening this goal. As Sanyal concluded, "they would rather see the urban poor either sent back to the rural areas" or settled in new cities.

La Revere (1986) described the programme of the Rio de Janeiro state electricity utility (LIGHT) that promotes urban agriculture on its right-of-ways. This is done to reduce maintenance costs (notably herbicide or manual weed control) and to increase food supplies for its employees as well as for local residents and schools. But its main purpose is to establish and maintain a good working relationship with the community in order to increase the chances of expanding transmission corridors without resorting to expensive underground lines.

Urban Agriculture

Urban agriculture provides a critical livelihood and source of food for many urban dwellers, particularly low-income households in developing countries. Urban agriculture, for the most part, is an informal activity practiced by poor and landless city dwellers. It usually refers to small areas, such as vacant plots, gardens, verges. balconies and containers within the city used for growing crops and raising small livestock or milk cows, for family consumption or for sale in neighbourhood markets.3 peri-urban agriculture refers to farming near cities, generally intensive semi- or fully commercial production. Based on a combination of national census data, household surveys and urban research projects in a number of cities, it is estimated that at least one quarter and in some cases a much bigger percentage urban and pen-urban households in developing countries are involved in agriculture.

Urban agriculture provides non-market access to food for low-income families, and also creates agricultural jobs and incomes. In places where rural transportation infrastructure is poor or where agricultural market systems are inadequate, both urban and pen-urban agriculture fill critical food supply gaps. Four major issues relating to sustainable development of urban agriculture retire attention:

1. First, land-use planning and management systems should be developed to programme the use of land for the most productive sustainable use, including provision of environmental services, such as buffer spaces, run-off management and erosion control;
2. Second, urban agriculture should be managed to minimize air and water pollution, including contamination of water supplies by nitrates and phosphates from fertilizers, and other undesirable environmental impacts;
3. Third, urban agricultural practices should be managed to protect food safety and health by preventing contamination of food with excess pesticide and herbicide residues, heavy metals, pathogenic organisms and other toxic materials; in particular, there are high health risks associated with the use of untreated or improperly treated sewage water to irrigate crops, including infection from pathogenic micro-organisms and helminthes (worms), such as *Ascaris* (nematodes) and *Ancylostorna* (hookworm). Recycled treated wastewater is a viable source of water for urban agriculture;
4. Fourth, efforts should be made to recycle urban wastes for use in agriculture, through waste handling and treatment facilities for sewage, organic solid wastes and other wastes. Such recycling could provide employment opportunities as well as useful materials.

Agriculture and Land Use in Peri-urban Zones

Four major issues relating to sustainable development of agriculture in peri-urban zones require attention:

1. First, conversion of land from agricultural to non-agricultural uses should be managed to promote the conversion of land with marginal agricultural productivity and preserve the best agricultural land for agriculture, taking into account the need to manage water use for agriculture and competing uses;
2. Second, advisory and support services should be provided to support a shift to sustainable intensive production for urban markets, and to develop agricultural processing enterprises and other non-farm employment opportunities;
3. Third, critical forest areas around cities should be protected and sustainable intensification of agriculture and agro-forestry should be promoted to meet urban demand for forest products and substitutes and reduce pressure on forest resources. This may require assessment of the economic, social and environmental costs and benefits of upward and outward growth of the city;
4. Fourth, changes in plant and animal species composition associated with changing land use and intensified cultivation in urban and peri-urban zones need to be assessed with respect to biodiversity. In the forest zone dramatic changes have occurred in tree cover, vegetative species composition and wildlife.

Rural Urban Linkages

Urbanization has a major impact on rural agriculture, including on what is grown, how it is grown and the sustainability of production systems. Rural-urban linkages depend on historical, political, economic, cultural and ecological factors, as well as on the macro-level policy changes being implemented.

At least six rural-urban linkages require attention in planning for sustainable agricultural development. One is the increased competition for natural resources due to growing urban demand and the resulting resource degradation, even in areas far from urban centers. Marketing opportunities in urban centers are contributing to an expansion of the area under cultivation in rural areas. Since land resources are limited. Increases in food supplies increasingly require productivity growth through the intensification of cultivation. Both the extension of cultivation into marginal lands and inappropriate intensification practices are contributing to resource degradation.

A second linkage is the role of urbanization as a catalyst for the commercialization of agriculture, which increases the use of fertilizers and other inputs, promotes rural non-farm enterprises, and increases urban-rural trade and other exchanges.

A third linkage is demand-induced commodity shifts in production. These commodity shifts, which lead to regional and sub regional specialization, could be among the most important factors affecting the sustainability of agriculture in the coming decades.

A fourth linkage is the impact of urbanization on rural labour supply. Urban industrialization and rural-urban migration may raise agricultural wages, increasing the demand for new agricultural technologies. Rural-urban migration has led to a "Youth drain" in many rural areas, with impacts on agricultural labour productivity and rural social life.

Reduced labour availability in rural areas, particularly during peak seasons. is encouraging the adoption of labour-saving technologies, including agro-chemicals and machinery, which increases labour productivity but also carries risks to land and water resources unless appropriate technologies are used. Moreover, poorer rural households that cannot easily invest in labour-saving technologies will face even greater problems unless there are compensating mechanisms available, such as labour-sharing and micro credit.

A fifth linkage is the food transportation, distribution and marketing system for getting food to urban consumers. There are many inefficiencies and service breakdowns in supply channels to and within cities. Wholesale and retail markets are often not well organized or managed, and market facilities are old, poorly maintained and too small to meet the needs of the growing population. Lack of efficient and reliable distribution channels imposes costs on urban consumers. There are also costs to the environment since more resources are used for production to make up for waste and losses in distribution systems.

A sixth linkage is the financial flows between urban centres and rural areas. Urban-rural remittances are one of the most important sources of investment capital in rural areas. Individuals returning from the cities following retirement or short-term employment bring savings, which are another major source of capital for agricultural investment. Remittances are important for the daily subsistence of some rural dwellers, such as many households headed by women. They are also important or acquiring durable non-investment goods and community social infrastructure.

In the other direction, resource flows from rural areas to urban centres have historically been a significant source of investment capital and stimulus to growth aid employment creation in urban centres, often at the cost of growth in rural areas. Policies and institutions affecting rural-urban resource flows will continue to have an impact on technological development and capital investment for both rural and urban agriculture.

Support for Sustainable Urban and Peri-urban Agriculture

Based on the importance of urban agriculture for household food security particularly for low-income households — and its impact on the welfare of women. Technical and financial support should be given to sustainable urban agriculture. Urban farmers need information about appropriate crops, fertilizers, and pest control and water conservation practices in order to improve productivity while reducing pollution and degradation. Institutions and regulations are also needed to manage the use of public and private land for urban food production. Organic production technologies and other farming techniques with low environmental impacts should be promoted.

For peri-urban zones, support should be provided for sustainable production practices for the production of commodities to meet urban market demands. For this purpose, there is a need to support and build on the adjustments that are being made by farmers to ensure sustainability under intensification. While not all adjustments are optimal, they reflect the perceived opportunities and constraints of farmers.

For both urban and peri-urban agriculture, there is a need to monitor assess the increasing competition between agricultural and non-agricultural uses land, water and other resources, and how such competition and associated changes in resource costs are affecting resource use, production systems and food supply, as well as social stability.

In relation to rural-urban linkages, there is a need to assess emerging patterns of land use and production in response to urban demand. Increases in distribution and marketing efficiency are needed at each link in the supply chain, including further development of agricultural processing activities and commercial institutions. Support should also be provided for rural industries that complement sustainable agriculture and land use.

Obstacles

In one of the rare observations on the attitudinal obstacles to overcome. Tricaud (1987) noted that there is a problem with the various specialists involved in urban land management. They tend to see only the single aspect of urban agriculture related to their own field and regard the other various functions as competing Interests.

Yet urban agriculture could fulfill most of their requirements if landscape architects, urban planners, economists, sanitary engineers, and agronomists were to co-ordinate their efforts and make a few compromises. The problem, of course, is that urban agriculture suffers from split-incentives and the lack of an integrated agency that can overcome these sectoral and temporal problems.

Wade (1987) emphasized the almost universal importance of government support in providing land and water sources in addition to an initial supply of planting material. Seeds and seedlings are often not readily available in the community, are too expensive, or are simply too old. A problem with supplies donated from outside is that they are often inappropriate for local growing conditions and/or cultural preferences. The most successful strategy has been collecting seeds from discarded marker produce.

Effective technical assistance, including soil testing, is also needed for urban agriculture. Unfortunately, many of the NGOs skilled in community organizing have little expertise in horticulture. As Streiffeler (1987) observed, urban agriculture cannot be considered as simply the continuation of the old habits of rural immigrants. There are fundamental differences between the extensive slash and burn agriculture often practiced in rural areas and the more intensive requirements of fixed farming in the city. Another drawback with urban agriculture is that the "survival margins" of urban farmers are so small that they cannot afford to experiment with new crops or techniques (Richards 1985).

Wade (1987) suggested numerous initiatives that could be used to promote urban agriculture and to improve access by the urban poor to other food supplies. These include granting tenure to squatters to encourage long-term investments in urban agriculture; requiring all new housing construction to provide adequate facilities for home gardening (including balconies and rooftops); and providing tax benefits to individuals or companies who "lease" their land for community food production.

What is needed is are-orientation of existing resources and programmes to include agricultural and food planning in urban areas. This implies more flexible land- use, water, and waste management. Production alone, however, cannot solve all of the problems: new efforts are also needed to improve food distribution, preservation, storage, and processing of food. Nutritional habits should also be improved by promoting the consumption of traditional fruits and vegetables. Efforts are needed to debunk" the myth that imported, Western produce is better than locally produced food (Wade 1987).

Priorities for Action

Priority areas for action include:

1. Strengthening capacity for land use planning in urban and peri-urban areas;
2. Designing and implementing regional development plans for integrated urban and rural development;
3. Strengthening extension and training services for urban, peri-urban and rural farmers in sustainable agricultural and resource management practices;
4. Improving food distribution, storage and marketing systems, and promoting the development of agro-processing activities to meet the needs of growing urban populations;
5. Improving the definition and administration of property rights to address conflicts over land and water use;
6. Strengthening the capacity of public administrations, farmers' organizations and other organizations to respond to new agricultural opportunities and problems resulting from urbanization through coordinated and cooperative efforts.

Discussing the Urban Agriculture, we find that the Government of India should promote sustainable development through many programs, policies and initiatives. For example, a key aspect of the Government of India's approach to sustainable development is the requirement for departments develop Sustainable Development Strategies (SDSs) and table them in Parliament every year. These strategies are important tools that help guide departments and agencies within the Government of India in systematically integrating the principles of sustainable development into their policies, programs, legislation and operations.

Conclusion

Part of the problem is that urban agriculture, at least in Africa, has simply "been developed to the point of enabling people to survive" (Streiffeler 1987). While critical of local authorities in Africa, he concluded that Western countries must also share the blame for their unwillingness to fund urban agriculture programmes.

It is clear that urban agriculture cannot replace other strategies (thus income redistribution. programmes must continue), but it has the advantage of generating independence. It uses many of the principles of self-reliant, local development based on initiatives that can be undertaken directly by local people using resources already available in the community. Urban agriculture also establishes direct links between the actions and outcomes while minimizing the risk of benefits being diverted to more powerful urban groups.

Gutman (1987) also noted that "urban agriculture helps to create and recreate the links of communication and action, of self-organization and initiative, that are absent in the traditional systems of income redistribution".

Nevertheless, one must plan for high failure rates, with slow growth levels and frequent abandonment's. Gutman (1987) estimated that it would take up to ten years for urban agriculture to reach 20 per cent of the households in greater Buenos Aires. This compares with the six months that it would take for a food distribution programme to reach the same number. It must be acknowledged, therefore, that urban agriculture is not a short-term solution and that it may not necessarily benefit the lowest income groups.

References

See United Nations, *World Urbanization Prospects: the 1996 Revision* (Sales No.E.98.X1II.6).

See United Nations, *an Urbanizing World: Global Report on Human Settlements 1996* (Oxford University Press, 1996). s

Urban agriculture also includes some intensive and well-managed production by commercial enterprises.

"Peri-urban "zones" are areas surrounding cities, which are in most ways integrated with the city. They extend as far as there is direct impact and influence of the urban centre, with greater population density, infrastructure and services than is generally found in rural areas. Depending on circumstances, a pen-urban zone can extend from the urban boundary to 50 or more kilometers from an urban centre. The issue of getting food into cities was identified and appraised as an emerging development priority in FAO document C.99/2, submitted to the FAO Conference at its thirteenth session, in November 1999.

Economic and Social Council, Distr. GENERAL 2 February 2000 ORIGINAL. Economic and Social Council Commission on Sustainable Development Eighth session 24 April-5 May 2000

Authors suggest their opinion.

29

Contribution of Marginal Women to Agriculture in India: Role of NGO's

Pavitra Ravindra Alur

Introduction

The women involved their roles and differing positions in society, however usually women are neglected in rural development even they experience equal status in the household decision-making process and are often described as the silent head of the home or "informal" power. Their roles in this societal context cover the spheres of human reproduction and equally important agricultural and household production, self-employment in the informal market sector and as wage laborers. The work of women and men differ yet the population is treated as one undifferentiated unit in some subsequent sections education, health and economic development. It is difficult to believe that women and men would play similar roles in economic development if their work ethic is so different. This reflected that women are not included in development planning. The reason why women excluded in development planning are they not given equivalent access to land, credit or extension services. Women also have potential to contribute to agricultural productivity beside the productivity of domestic activity is another extremely important area which should not be ignored by planners. Gender issues in development are a relatively new area of research of much importance because of its potential impact on shaping the societies of developing countries.

Indonesia is in a good position to integrate rural women into development because social values in its cultures such as the Javanese already provide them with relatively egalitarian status. The lack of consideration, however, for women in the development literature reflects a need for development officials to start including them with the goal of development being one that benefit to the whole rural community. There is evidence to indicate that by eliminating barriers to women's access to productive assets, they can fully participate and be recognized as important partners in the development process.

Women at work in Sharanasirasagi (Dist: Gulbarga).

It is a historical fact that there has been a long, long agriculture has taken for its development. For example man started cultivating land with the help of another man and both of them doing manual labor. Similarly as the age passed by, many changes have come into place in the cultivation of land. The modern agriculture of 17th -20th century has witnessed a sea – saw change.

In the cultivational practices. In the initial years man started using wild animals for the cultivation of land. Further he also gradually shifted to horses, he buffalo & specially in India switched on to bullocks. He also started using the iron ploughs. In the modern days farmers are using tractors for cultivation.

With this background, if we turn to use of labor, there has been a substantial change in agriculture. Not only women were considered as gatherers of the bounty of hunting and cooking the food. Where as in Indian agriculture there is no exact date to mention that when & where women become the equal partners in the process of cultivation of land.

Further more not only women are farmers but also agricultural laborers. There, the contribution made by women is difficult to evaluate. Therefore the present paper attempts to explain the contribution of modern women in the cultivation of land in India in the recent years. For convenience the paper is divided into following parts.

- The Part I briefly explains the growth of female agriculture labor in India.
- The part II provides a picture of activities in which they are engaged in modern agricultural practices.
- The part III reveals the problems faced by marginal women in agriculture.
- The Part IV is about suggestions for improving their welfare.
- The Part V Summary & Conclusion.

Women cutting crops at Shahapur (Dist: Gulbarga)

Part I: Growth of Female Agriculture Labour in India

During last few decades (After independence) there has been tremendous growth of population in India. As a result, there has been great dependence on agriculture for the purpose of survival. This does not mean there is zero productivity of this labor.

The initial decades of development have brought a transactional change in the growth of population. This change has been growing at a fast during the last 2 decades.

The female work force participation rate in rural areas in much higher than the urban areas.

1. **Variation in the women labor participation**: Along with the many factors of changing trend visualising in the economy there became regional variation in female labor force participation. The variations can be studied in the following ways.

   (a) ***Very high female work Participation***: The states of North eastern region like Nagaland, Arunachal Pradesh, Manipur, Meghalaya, Mizoram and Sikkim show a very high female work participation rate.

Women sowing seeds at Shahapur (Dist: Gulbarga)

2. **High female work participation**: The southern or south western states like Andhra Pradesh, Maharashtra, Madhya Pradesh, Tamil Nadu and Karnataka came next in order.
3. **Low female work participation**: Kerala which has a very high level of literacy (86.2% 1991census) has a female work participation rate of 13% which is below the all India average.
4. **Very row female work Participation**: Similarly some of the economically better off states such as Haryana, Punjab and west Bengal show a very poor female participation.

The types of participation in which women are found indicate the low level of their participation in the economy. While the representation is reliably high in agriculture labor and household industry, women constitute less than 20 per cent of the work force in the other categories. It is quite evident that there is need for a substantial books in women's employment in other sectors.

Part II: Modern Agricultural Practices

Indian agriculture has witnessed a slow and steady growth since independence. The average growth rates varied from 2.9 per cent during the first plan (1951-56) to 4.3 per cent during seventh (1985-90). The compound growth rate of agriculture output has been around 2.7 per cent annum during 1949-50 to 1988-89. It has increased at a rate of 3.1 per annum during the pre-green revolution period (1949-50 to 1964-65). During this period, area under cultivation expanded at an annual rate of 1.6 per cent which was the pre dominant factor explaining this growth. The growth rate of agriculture production in the post-green revolution period was 2.7 per cent per annum. However the increase in productivity turned out to be the major contributory factor during this period. The break – through in teens of seed fertilizer irrigation technology led to increase in the growth in productivity to 2.0 per cent in the post-green revolution period as compared to 1.3 per cent in the post-green revolution period. Two striking features of the agriculture growth pattern name, however, been. (a) instability in year to year production (b) inter-regional and inter-crop disparities in production performance.

As the modern technology of agriculture has been mostly irrigation-dependent, its gains remained confined mainly to areas / region favorably endowed in expect of irrigation. During the 80's the growth rate of agricultural production was lower at 2.4 per cent, despite special emphasis on the supply of inputs. Indian agriculture needs a new thrust and attention in the 90's.

Post-independence decades have witness a valuable transformation in Indian agriculture. What is most readily noticed is the increase in agriculture production.

Particularly the nearly four-ford rise in the production of food grain which saved the country from on of the greatest crisis it faced after independence. The increase in production was made possible by expansion of irrigation, spread of new technology and additions to rural infrastructure like irrigation works, road, market yards and ware houses, These changes in the physical features of agriculture needed changes in the institutional structure. Like new credit institution, extension agencies, regulated markets, control of money lenders. The cumulative effect of these changes has been reflected in the formers their perceptions attitudes & behaviour and in the changing features of rural communities such as satisfaction, diversification commercialization exposure to outside influences. The total process bringing about these wide ranging changes could be described as modernization of agriculture. In a rural cum agriculture based society like India, the process of modernization of agriculture through its direct & indirect impacts would be a major determinants of economic status, development and welfare of rural people.

A striking feature of modernization of agriculture in India is the contrasting scenarios through which it is unfolding. The aggregative physical parameters of agriculture show remarkable achievement. Listed below are a few striking indicators of improvements in the physical aspects of agriculture.

1. Between 1949-50 and 1996-97, Index (base 1980-81) of agricultural production increased from 49 to 176, index of food grains from 52 to 161

Women Plucking and cleaning red chillies

Women planting the plants

Women cleaning crops

and non-food grains from 45 to 201. Index of per hectare yield (all crops) increased from 74 to 149. It may be mentioned that the agricultural scenario has been one of stagnancy in the decades prior to Independence.

2. Between 1950-51 and 1994-95, gross cropped area increased from 132 million hectares to 188 million hectares implying a rise in cropping intensity from 111 to 132. Gross irrigated area went up from 23 million hectares to 71 million hectares.
3. Between 1950-51 and 1996-97, distribution of improved seeds increased from negligible to 70 lakh quintals. Fertilizer use increased from only 69 thousand tones to 14 million tones with per hectare fertilizer use going up from negligible 10 to 77 Kgs. Pesticides use increased from 2.4 thousand tones to 56 thousand tones. Consumption of electricity in agriculture went up from 15200 Million KWH in 1981-82 to 85736 Million KWH in 1995-96.
4. Institutional credit (Cooperative Societies, Commercial Banks and Regional Rural banks) was 28653 crores in 1996-97 as compared to 10186 crores in 1989-90.

Sector-wise distribution of women

Sectors	Rural India			Urban India		
	1972 -73	1983-84	1990-91	1972-73	1983-84	1990-91
Primary (Agr. Sector)	90	88	85	34	32	25
Secondary (Non-farm sector)	6	7	8	28	30	31
Tertiary (Non-farm sector)	4	5	7	38	38	44
Total	**100**	**100**	**100**	**100**	**100**	**100**

Part III: Problems Faced by Marginal Women in Agriculture

In the outgoing century more theoretical discussions regarding women labor, took place including gender-based distinctions to justify discriminations and home economics with household production etc. All these were based on gender differences. Women laborers constitute significant portions of the total labor force. However, labor markets do not play fair to them. Their labor does not receive expected attention and sometimes, becomes marginalized in the process of development.

1. Generally women laborers are not treated equally in the labor market.
2. They have no equal access; to the contrary there is far difference in them.

3. Women are predominantly employed in the labor intensive unskilled jobs in the unorganized sector with less security of facilities for partial tenure and lower wages, while their male counterparts have access to capital-intensive skilled jobs in the organized sector.
4. Women are paid less (payment) than men for the same nature of jobs.
5. Men may not be able to earn an income which will fulfill family requirements.
6. **Seasonality**: Though agriculture is a seasonal activity, the women laborers will sit without any work in Non-reason days, so, it's is very difficult in making life round the year.
7. **Poor facilities**: Women are not provided any facilities during work.
8. **No security**: There is no security against risky work.
9. Heavy burden of work
10. Poor nutrition
11. Lack of leisure time

To, these add multiple pregnancies at short intervals, a big family to cater for poor health services, water supplies, and housing which are her daily lot, and the life of a rural women becomes almost unbearable.

Part IV: Suggestions and Improving their Welfare

1. The wages should be increase
2. Providing continuous work
3. They should be provided family benefits
4. Separate ways for risk and difficult work
5. Bank should give loan without interest. They should be given that relief.
6. Insurance courage to women
7. National rural employment guarantee act (NREGA). Should be implemented successfully.

Part V: Summary & Conclusion

Through local initiative, organizations large and small from the village to the national level have been started for the purpose of improving the socio-economic status of women. To improve agriculture women's projects there is a need for introduction of labor saving farm machinery and equipment provision of agro-chemicals, improved seeds, fertilizers and weed killers at affordable prices.

The role of NGO is as follows:

1. Education should be provided to the women laborers by the NGO's
2. Skill development training should be given them by the NGO's.
3. Social workers should take an initiation to organize women laborers and strengthen their group.

References

"Economic Planning and Policy", Dr. S.S.M. Desai (1986) Himalaya Publishing House, Bombay 400004.

"Unlimited Supply of Labor", Arthur Lewis (1968).

"Women and Employment", Dr. (Mrs) S. Murty (2001) RBAS Publishers Jaipur.

"Rural Development Issues", Dr. D. Vasudev Rao (1988) ESS ESS Publications New Delhi 110002.

National Sample Survey Organization (1991).

Agricultural Statistics at a Glance, Government of India, March 1998.

Index